German Annual of Spatial Research and Policy

Editors:
ARL – Academy for Spatial Research and Planning, Hannover
BBR – Federal Office for Building and Regional Planning, Bonn
IfL – Leibniz Institute for Regional Geography, Leipzig
IOER – Leibniz Institute of Ecological and Regional Development, Dresden
IRS – Leibniz Institute for Regional Development and Structural Planning, Erkner (near Berlin)

Responsible for the 2007 edition:

ARL – Academy for Spatial Research and Planning

Dietmar Scholich (Ed.)

Territorial Cohesion

With 18 Figures

Springer

Editor:

Prof. Dr.-Ing. Dietmar Scholich

Academy for Spatial Research and Planning
Hohenzollernstr. 11
30161 Hannover, Germany

In the German Annual are published scientific articles, which were reviewed according to international standards, as well as short reports about research work and practical experience.

Editorial staff:

ts redaktion, Berlin: Thies Schröder, Vera Hertlein, André Eißer

English language editor:

David Skogley

Layout:

André Eißer, Berlin

Library of Congress Control Number: 2007941256

ISBN 978-3-540-71745-4 Springer Berlin Heidelberg New York

Springer is a part of Springer Science+Business Media

springer.com

Typesetting: by the editors using a Springer TEX macro package
Production: LE-TEX Jelonek, Schmidt & Vöckler GbR, Leipzig, Germany
Cover design: eStudio Calamar S.L., F. Steinen-Broo, Girona, Spain

SPIN 11936176 7/3180/YL - 5 4 3 2 1 0 Printed on acid-free paper

Acknowledgement

Special thanks to the following authors: Joachim Burdack, Hans-Joachim Bürkner, Sylwia Czarnecka-Zawada, Carl-Heinz David, Maroš Finka, Hans-Peter Gatzweiler, Markus Hesse, Gerold Janssen, Rupert Kawka, Robert Knippschild, Beate Konienczny, Markus Leibenath, Ulf Matthiesen, Kristine Müller, Eva Reisinger, Lars Stratmann, Wendelin Strubelt and Sabine Zillmer for their contributions to this publication. I would also like to thank Heiderose Kilper (IRS), Sebstian Lentz (IfL), Bernhard Müller (IOER), and Wendelin Strubelt (BBR) who have also supported this publication as editors. Further thanks go to André Eisser from ts redaktion in Berlin for his competent editing of our annual.

Dietmar Scholich

Contents

Short Reports

1 Preface – Introduction

Dietmar Scholich

The various regions within the European Union are marked by a high degree of disparity with regard to economic performance and productivity, and with respect to their labour markets. The intention underlying the EU's concept of *'territorial cohesion'* is to reduce these differences and imbalances and, at the same time, to ensure more balanced spatial development throughout the EU. Territorial cohesion – for which the German language uses two slightly different terms, "territoriale Kohäsion" and "territorialer Zusammenhalt" – has to be seen as a general term encompassing a number of different components. In addition to redressing disparities and spatial imbalances, this concept aims to improve territorial integration and to promote co-operation among EU regions. In striving to bring about territorial cohesion, due account must be taken of the diversity of the European continent, and consideration should be given to the advantages and disadvantages each area exhibits. Among the special challenges to be faced are the concentration of population and economic activity in what has come to be known as the Pentagon, the process of EU enlargement, and the severe imbalances which exist between the major centres of population and other parts of the EU territory. Against this background, it was only right and proper that territorial cohesion should be included in the draft of the Treaty establishing a Constitution for Europe (Article I-3), which lists economic, social, and territorial cohesion as being among the Union's objectives.

Nevertheless, to date there is still no general consensus in professional circles in Germany on how the term (or terms) is to be understood, nor has any one definition of *'territorial cohesion'* established itself – even though the concept has been in use since the mid-1990s, when it appeared in the European Spatial Development Perspective (ESDP), and has since been mentioned in both the second and the third cohesion reports produced by the European Commission. The EC Treaty makes explicit reference to territorial cohesion solely in Article 16 in the context of services of general economic interest in the shared values of the Union. This provision guarantees services of general economic interest (i.e., access on the part of EU citizens to services, basic infrastructure, and knowledge) and underlines their influence on the common values of the Union as well as on social and territorial cohesion.

Territorial cohesion as an EU objective and task

If we regard territorial cohesion in its broader sense as the balanced spatial development of the European Union, then it is a fundamental pillar of sustainable development; indeed, it constitutes the implementation of the spatial dimension of this concept. In this respect, territorial cohesion – even in the absence of any specific reference to the term – can be subsumed under the EU objectives contained in Article 2 of the Treaty on European Union. These objectives include the aim to *"promote economic and social progress and a high level of employment and to achieve balanced and sustainable development, in particular through the creation of an area without internal frontiers, through the strengthening of economic and social cohesion and through the establishment of economic and monetary union, ultimately including a single currency in accordance with the provisions of the Treaty."* In as much as achieving territorial cohesion is viewed as a contribution to redressing disparities and promoting integration at the spatial level, it represents an essential complement to social and economic cohesion in the EU.

As the third pillar alongside social and economic cohesion, territorial cohesion also represents an EU-level task. Accordingly, territorial cohesion was recognised as early as the second cohesion report of being an integral component of regional policy under Community law which needed to be strengthened in order to promote the *"overall harmonious development of the Community"* (Article 158 EC Treaty), even if no explicit reference to it was included here either.

Instruments to secure territorial cohesion

Viewed from the perspective of spatial planning research, territorial cohesion relates to a series of EU policies, and consequently calls for a new quality of integration with regard to spatial development in Europe as one necessary condition for sustainability and competitiveness in a globalised world.

Aspects of territorial cohesion have a particular impact on the regional policy of the European Community. The instrumental means for implementing this policy are provided essentially through grants. Various funds – such as the Structural Fund, the European Fund for Regional Development, and the Cohesion Fund – have been created specifically to strengthen economic, social and – implicitly – territorial cohesion. One instrument specifically tailored to the objective of territorial cohesion is the EU's INTERREG programme. This is a Community initiative introduced within the framework of the European Fund for Regional Development as a means of promoting co-operation among EU regions. What the INTERREG programme seeks to achieve in concrete terms is to institute cross-border (INTERREG III A), transnational (INTERREG III B) and interregional (INTERREG III C) co-operation within the EU. Special attention is given here to the most peripheral regions, as well as to those abutting prospective new member states. The INTERREG programme is also supported by ESPON, the European Spatial Observation Network. This Community initiative was originally scheduled to run from 2000 to 2006. The legal package for the 2007 to 2013 funding period, however, continues to recognise territorial cohesion in Europe as a Community priority (Objective 3), which has led to the INTERREG III B programme on transnational co-operation being extended to run through the 2007 to 2013 funding period.

From territorial cohesion to a European competence for spatial planning?

An issue closely related to the concept of *territorial cohesion* is the question as to whether responsibility for spatial planning and spatial development policy should be located at the European level. Although it is true that efforts to achieve territorial cohesion do address concerns of direct relevance to spatial planning, there are at present insufficient grounds to justify the creation of an independent area of competence for spatial planning. To date no member states have shown any signs of ceding their responsibility for spatial planning.

Since spatial planning and spatial development do not have the status of independent areas of competence, spatial development can only ever be addressed through measures originating in other policy areas, for example – as mentioned above – within the framework of regional policy, or also in the process of creating or extending trans-European networks, and in the context of activities associated with transport, agricultural, research, technology, environmental and competition policy. Sectoral policies, however, are in many cases each based on their own distinct understanding of the concept of space and thus take little or no account of other concerns affecting spatial development. They thus fail to give adequate consideration to the objective of achieving balanced spatial development.

It is also doubtful whether the EU Constitution will do anything to change this situation by explicitly enshrining the concept of territorial cohesion and thus introducing a specific competence for European spatial development policy. The arrangements which have so far been included in the Constitutional Treaty would suggest that this is not likely. The approach adopted is rather that of setting the course for EU policy and action, and equally of creating the single European market; consideration is to be given to achieving the objective of territorial cohesion. What this means is that the individual sectoral policies are required to be co-ordinated in dealing with any matters which have spatial impacts. Whilst this approach entails a mandatory requirement across policy areas to take the objective of achieving territorial cohesion into account, it does not provide a basis for establishing a distinct area of competence for measures which only concern spatial development.

Implementation of the objective of territorial cohesion calls for knowledge of the conditions and circumstances affecting spatial development in and between European regions, as well as knowledge of the spatial impacts of EU sectoral policies and a Community-oriented interest in spatial development. Despite ESPON, there are still major deficits here; consequently, the role of spatial planning at the European level needs to be reinforced. Moreover, there are major differences between the systems and philosophies underpinning spatial planning in the various member states – and apparently no signs of any significant harmonisation. It would be extremely helpful if the member states could even just agree on certain minimum standards of spatial planning policy.

Since the introduction of the term *territorial cohesion*, a wide-ranging discussion has been in progress on the content, character, and value-added of this concept. So far it has not been possible either to resolve the terminological problem affecting the parallel German terms or to overcome the uncertainty which has surrounded the interpretation of the concept. The controversy in this debate has been both over political and legal interpretations of the term and on the expected effects and side-effects. This has been accompanied by an unmistakable lack of interest on the part of the European Commission.

In bringing together contributions from their respective institutions, the editors of the Annual – representing Germany's principle institutions for the spatial sciences – are hoping to give the discussion on territorial cohesion in Europe renewed impetus. This theoretical concept has the potential not only to improve both the horizontal and vertical co-ordination of the complex technical demands that apply to the development of European regions, but also to smooth the path in the drive towards sustainable spatial development and equivalent living conditions throughout Europe – if it proves possible to transfer this theoretical construct into a political concept of territorial cohesion.

2 Status and Perspectives of Territorial Cohesion with Respect to European Spatial Development Policy (ESDP) - Normative and Governance Aspects

Carl-Heinz David [1]

Introduction

Treating 'Territorial Cohesion' (TC) is similar to Plato's cave parable: *you are watching phenomena, like shadows on the wall of a cave, which could be interpreted in one way or another, but you cannot perceive the reality hidden from you which causes these phenomena on the wall.* Contrary to political demands for more transparency in European policies and more opportunities for the involvement of citizens, the insight into TC policy procedures appears to be widely secluded from the public and even hidden from research interests, and perhaps from many official experts involved in the debates as well. When lengthy procedures have come to an (interim) end and formal, political decisions have materialized in final (conference) documents which are submitted to the public, and are hence available to the scientific community, it is too late in most cases for any public or scientific involvement in the procedure and its results.

Due to EU finances – e.g. ESPON projects – there is a lot of research, but attributions to the transparency of the scientific and political implications are restricted. Politically, the EU channels the results, in particular by financing only limited research mandates, through influencing the time of publication of the results, and by using the research findings in a politically selective manner. For example, comparing studies on the governmental systems of various member states sometimes provides a rather simplified ('standardized') view of the constitutional, administrational and legal internal structures of these member states, but rarely illustrates how the systems work in their complexity, or provides a more sophisticated base for comparative considerations. Research information has increasingly become an important policy control tool that the EU gets involved in and then politically tries to control areas of competence and administration normally managed by the member states and not primarily by the EU. In particular, territorial cohesion-related political areas such as regional/spatial planning and urban policy are an example of this problem.

Even the sophisticated research now available in the wide ocean of internet information sometimes produces only irrelevant and outdated findings. The access to and freedom of information does not apply to EU practices. Many freely distributed information sheets and brochures only allow for a politically channelled,

[1] in collaboration with Hendrik David (Dipl.Arch. ETH/MA HSG)

sometimes rather outdated view. It is the same with relevant documents with respect to the current *Status and Perspectives of Territorial Cohesion* in the EU, to those which are cited, mentioned and referred to in this paper. Existing deficiencies in information can only be balanced based on respective presumptions or even on speculations in many cases.

2.1 Main Sources of Information

The European Treaties and the European Constitution as a primary basis for TC

The primary basis for normative and governance aspects of TC are the European Treaties. On the one hand there is the important Treaty of the European Community (TEC).[2] On the other hand there is the Draft of a Treaty Establishing a Constitution, 2003 (DraftTCE).[3] The latter might have been useful for prognostic purposes but must be considered as having currently failed.[4] The TCE nevertheless provides information about what the probable content of a future European Constitution based on a new or modified approach shall be. Specifically referring to the aspect of spatial development policy, a document of great relevance is the European Spatial Development Policy (ESDP).[5] This programme is currently in a transitional stage and will be expire soon and then be replaced by a new policy approach. This stage is often called the post-ESDP-period. A pure amendment of the ESDP is not to be expected. Decisions have been made, however, which now assure a certain continuity with respect to the future development of the EU territory. A political 'road map' of the EU Cohesion/Territorial Cohesion Policy ('Territorial Agenda')[6] is planned. It

[2] European Community (2001)

[3] European Convent (2003) Draft of a Treaty Establishing a Constitution for Europe, submitted by the European Convent to the European Council in Rome in 2003 (2003/C 169/01) (DraftTCE)

[4] Due do to a recent EC Council resolution (6/2006), the German Presidency has been commissioned to submit a report about how to revive the European Constitutional Procedure in 2007. This resulted in the Berlin Decleration on the occasion of the 50th anniversary of the Treaty of Rome in March 2007 looking forward 'to a new legal base' to be effective before the next European Election 2009. The new Portuguese Presidency already submitted 'Draft Treaty' documents amending the TEC to the Intergovernmental Conference (IGC) in charge of the treaty amendment on 24/07/2007. This reflects the new terminology applied avoiding any designation as constitution. In respect to the TC issue the new draft documents appear to take over the respective articles of the DraftTCE without major deviation.

[5] European Spatial Development Policy *(ESDP),* finally adopted at the Potsdam Conference in 1999.

[6] See Territorial Agenda of the EU, Guidance Note as of 26/06/2006, agreed upon at the Informal Ministerial for TC held in Leipzig on 25 May 2007. In reference to the separate document "Territorial State and Perspectives of the EU;" see below, footnotes 12 and 13.

was presented to the public in May 2007 during the German EU-Presidency. In this paper political definitions with respect to further Territorial Cohesion were given. So far, as the revision of the ESDP has to be implemented within the political realm of the Directorate General (DG) XVI of the EU Commission, it is legally based on its competence for the regional/cohesion policy. At the same time this policy has elements of a sectoral task, as it is connected to the complex regional/coherence fund system. But DG XVI politically is trying to extend this sectoral approach to a cross-sectoral political approach. Regional/spatial policy of DG XVI is primarily restricted to sectoral fund management, as the competence for a comprehensive spatial development policy remains, according to Art. 5, par. 1 TEC, with the Member States. Hence, the competence for spatial planning is not explicitly transferred to the Community. Various efforts, however, have been made by the Commission/ DG XVI to overcome these restrictions and to extend EC competence to respective areas of competences in the realm of the Member States.

The legal base lies in a constitutional contradiction which, according to Art. 5 TEC, states that the EC is restricted to the competence explicitly attributed to it; but the EU/EC treaties provide clear legally defined areas of competence, orientate them with respect to the goals to be achieved by the EU, supported by emergency competence regulations.[7]

In the past the ESDP was worked out *'intergovernmentally'*, but with participation and financial support by the Commission (DG XVI). As the EU Commission felt that it had no control of managing functions, the Commission (DG XVI) discontinued this cooperation, referring to reasons of 'comitology'. Within its proper competence the Commission (DG XVI) has prepared an ESDP discussion paper[8] supported by national experts.

Other policy instruments of the Commission involve the preparation and publication of reports on cohesion policy, of which the 3rd Report on Economic and Social Cohesion[9] is currently the most relevant, because it also deals with aspects of TC. A key role in the post-ESDP-process is played by a document which continues to be a subject of discussion.[10] There are many versions of this, reflecting various disputes, agreements and unanswered questions between the (Informal European

[7] David (1993) pp 1021-1029

[8] Working Group on Spatial and Urban Development (SUD) (2003) Managing the Territorial Dimension of EU Policies after Enlargement, 1/9.2003

[9] ECE (2004)

[10] The preceding paper was the so-called "Mermaid Paper" (a result of a former Stockholm Conference). The present paper goes back to the conference of the above-mentioned Informal European Council, which is sometimes called the 'Informal Ministerial Meeting on Regional Policy and Teritorial Cohesion', which took place on 29 November 2004 in Rotterdam and 20/21 May 2005 in Luxembourg. 'Informal' hints at the legal implication, i.e. the subjects discussed by the ministers (with attendance by the EU Commission) do not or at least do not totally belong within the EU's realm of competence, but are primarily within member states' realm of competence.

Council of) Ministers for regional/spatial Development and the EU Commission, represented in particular through the DG XVI. This document is called 'Territorial State and Perspectives of the European Union – Towards a Stronger European Territorial Cohesion in the Light of the Lisbon and Gothenburg Ambitions', hereafter referred to and cited as TSPDoc. There are various versions of this paper and it will be the base for the 'Territorial Agenda of the EU'. It is currently incomplete and in the phase of being edited. The permanent status of the edition is not available free of charge on the internet because there are discussions between the national ministers, Member State governments and the Commission.

The 2005 version (hereafter cited as TSPDoc 2005) has been widely available to general researchers[11] and a more recent extended version is currently (September 2006) available, hereafter cited as TSPDoc 2006.[12] Although the general structure appears to be the same to a certain degree, there are nonetheless huge differences. The contents were more than doubled, to about 70 pages. A still incomplete framework of priorities and sub-priorities were developed. It appears as if the *'stakeholder' policy* approach has been intensified in order to get control of spatial development-relevant sectoral actors on both the member states' and EU levels.[13] The shorter 2005 version is in many respects less confusing than the 2006 version, in particular as it provides a summary. The 2006 version is a more confusing than enlightening document, which has three organisational levels: parts, chapters and margin/text numbers and a table of contents without page references. Politically-speaking, the 2006 version certainly allows readers to draw a wide range of political conclusions from it and thus to reach a political consensus on it. In a way, the document contrasts with recently published, comparatively rigidly formulated EU-Regional Fund distribution guidelines, which otherwise do not reflect any influence derived from TSP documents.

---

[11] It was submitted by the (Informal) EU Council to the European Conference of Ministers responsible for regional/spatial planning – CEMAT and was published as CEMAT document CHF 83(2005) 14 (English only), although it is in substance of course an EU-Council Document.

[12] First Draft presented to all 25 EU member states on 6 June 2006, further developed by the Coming Presidencies Group - CPG/Editorial Group, to be presented to the CPG-meeting on 21 June 2006.The Territorial Agenda of the EU has been finally agreed on the occasion of the Informal Ministerial Meeting on Urban Development and Territorial Cohesion in Leipzig on 24/25 May 2007, based on an updated version of the above mentioned political background paper "Territorial State and Perspectives of the EU- TSPDoc , see http://www.bmvbs.de/en/Spatial-development-1876.963636/Territorial-Agenda-of-the-EU.htm

[13] An EU Stakeholder Conference under the Rotterdam Agenda on territorial cohesion with the title: *Territorial Cohesion and the Lisbon Strategy: Exploiting Europe's Territorial Potentials*, Amsterdam, tried to get 'stakeholders' cooperation in June 2006.

The various levels of European spatial policy

It is necessary to distinguish between the various levels of European spatial policy. There is an older 'layer' based on a European Council legal basis. This will be widely disregarded here, as the respective CEMAT (European Conference of Ministers Responsible for Regional/Spatial Planning) primarily and politically concentrates on dealing with aspects of sustainability. Territorial Cohesion, however, is dealt with within the realm of member states through informal European Councils, held by the (EU) Ministers for Spatial Development under the EC/EU treaties. This informality derives from the fact that Spatial Planning was not within the EU's original realm of competence. CEMAT brings together a wide range of ministers (including non-EU, but European Council member states) who are responsible for regional/spatial planning. Concerning this, we will hereafter concentrate mainly on EC/EU law, organisation, and administration.

Within the EU/EC's legal and administrative structure, member states politically interact in two very different ways: within the organisational and procedural frameworks provided by the EU/EC treaties and its implementation by the Commission (e.g. in EU Commission- controlled sub-commissions etc.), and in areas of proper member state competence with more or less reference to EU/EC activities, independently in a so-called *'intergovernmental'* manner. Spatial planning is one of the member states' political activities, where they have the choice to get involved in terms of competence in one way or another, and where, respectively, the Commission/the specific DG in charge has to politically respond to this componential *'weakness'*, as do other EU bodies (the European Parliament) and entities (the Committee of the Regions) as well.

The ESDP is, for example, an approach by member states and the Commission, where inter-governmental and the genuine EU/EC competence of the Commission were interrelated and brought to political decision making. Concerning various reasons discussed hereafter in more detail, this approach was discontinued by the Commission and therefore new organisational and procedural measures for TC are needed in the post-ESDP-period. A large change is due to the fact that organisational and procedural co-operation between the Commission and member states, as earlier applied for during preparation of the ESDP, has substantially changed since the final adoption of the ESDP in 1999 (see below, 2.4). The distribution of power and competence between the EU and member states is only one of various issues and not even the key aspect when the governance issue is raised. Governance covers a wide range of aspects, as European policy[14] has to be run and implemented in a complex multi-level governance system. There are various EU-mandated research approaches which provide a comparative view of the situation, but politically and scientifically it is not clear what standardised consequences have to be derived from the stated multi-level structure.

[14] ECE (2001)

2.2 Terminology applied

Territorial cohesion within the trilogy of economic, social, and territorial cohesion

The terminology used in official documents and in various research papers is not at all consistent. Only selected aspects can be focused upon here. Obviously, territorial cohesion is just one aspect of the terminological trilogy of economic, social, and territorial cohesion which is addressed in Art. I 13, par. 2 DraftTCE as an area of (future) shared EU competence. It is difficult to say whether the territorial aspect is adding something new to economic and social cohesion or whether it is already contained in these two aspects. The TSPDoc[15] rather indefinitely designates TC as an integral part of economic and social cohesion. Therefore TC cannot be interpreted by its wording alone. There is a need for additional consideration of the legal background of the distribution of competence.

As mentioned before, the competence issue is rather complex. The distribution of power between the Commission and member states is just one aspect. The issue is further related to inner-constitutional structures: with respect to other European institutions, e.g. the European Parliament (EP), the member states' federal/central state structures and, additionally, even to that of their inner-governmental structures related to the various ministerial departments involved, and their regional and local structures etc. This complexity has to be taken into account, even though it is in many respects not under direct EC law control and influence.

Territorial cohesion policy - objective or instrument?

A further complication arises from the biased character of TC as a policy objective and as an instrument or a policy tool. It is not uncommon in planning that, depending on consideration of the various levels of legislation, government and administration objectives and tools change their quality.

TC as a policy objective has to be considered as depending on higher ranking political objectives, like the following:

- *'Lisbon and Gothenburg Ambitions'* with a territorial dimension, such as:
    - Exploiting the endogenous potential of an area, including natural and cultural values
    - Promoting an area's integration and connectivity to other areas
    - Territorial governance: promoting horizontal and vertical coherence

- The fundamental EU goals of *'economic and social cohesion, conservation of natural resources and more balanced competitiveness of the EU territory in a territorial setting'*[16]

---

[15] TSPDoc (2005) p 4

[16] ESDP (1999)

- The ESDP principles:
    - Development of a balanced polycentric urban system and new urban partnerships
    - Securing parity of access to infrastructure and knowledge
    - Sustainable development, prudent management
    - Protection of nature and cultural heritage.

From this set of higher ranking objectives and goals, three specific strategic policy objectives for strengthening TC are identified by the EU Ministers for Spatial Development and the Commission,[17] which have to be translated into more specific priorities:

- Improving the strength and diversity of urban centres/networks as motors for territorial development
- Improving accessibility and territorial integration
- Preserving and developing the quality and safety of Europe's natural and cultural values/developing sustainable urban-rural linkages,
- All with special regard to strengthening the territorial capital of areas with weak economic structure/physical or geographical handicaps

The instrumental aspect of TC covers the wide range of territory-related (spatial or territorial) development policies. They are only within the realm of the ministers for spatial planning to a certain extent, but are to be implemented by various political 'stakeholders'.

Member states | commission | stakeholders

TC reflects the difficulty of coping with the political intention to co-ordinate/control without having the respective legal competence. In this situation there can only be an appeal to voluntary subjection of the addresses, and the EU tries to achieve this through budgetary incentives. The TC conceptually follows these control potentials. The strategy of the future EU Cohesion Policy will play a key role within the European Territorial Cooperation (ETC) strand.[18] This implicates an informal approach of bringing existing stakeholders in line with TC based on analyses and priorities of elaborated assessments of *'territorial state and perspectives of the EU'*, which are considered to be finally adopted in the concentrated form of *'Territorial Agenda'* mentioned above.

The member states' ministers who are responsible in their respective countries will meet in Informal Ministerial Meetings on Regional Policy and Territoiral Cohesion, as they have done in May 2007 in Leipzig. They are well aware of the fact that they are not masters of the *"Territorial Agenda"* procedure. They do not have

[17] TSPDoc (2005) p12

[18] TSPDoc (2005) p 21; TSPDoc (2006) p 74

a "formal obligation or incentive to take the territorial impact in account into the EU policy process."[19] They rely on a so-called *'evidence-based'* approach and the aforementioned *'stakeholder approach'*, with respect to their internal governmental structures and to the EU organisational structure. The Commission was asked to create an accountable contact point for territorial cohesion and for integrating the territorial impact into the EU policy development. This could be implemented through an impact assessment procedure and inter-service consultation (which refers to political co-ordination of the various DGs involved). This obviously reminds one very much of the strategies applied in the past to establish controlling powers for EU environmental policy.

Territorial cohesion – European spatial development policy

To a certain degree, TC is taking over the role of spatial development policy in the post-ESDP-process. In respect to the ambiguous distribution of powers and competence in the relevant fields of politics within the realm of the EU and member states, both TC and ESDP illustrate the same feature, i.e. that the borderline between both areas of competence is difficult or even impossible to draw due to their overlapping and competing structures. The unsolved overlapping of powers and competence (including a shared exercise, which does not precisely say where to draw the borderline) applies as well to TC.

2.3 Expanding Territorial Cohesion in the constitutional Debate

Current legislation: Art. 16 TEC

The terminology of *'territorial cohesion'* is already contained in Art. 16 TEC (Nice version)[20], but in a slightly different context, as it is applied in Art. I 13, par. 2 DraftTCE. Under the current EU legislation the aforementioned Art. 16 TEC applies to a specific aspect of EU policy: services of general economic interest.[21] This regulation has gained great political importance in some member states, such as France, whereas in other countries, for instance in Germany, this regulation has never become a large issue of spatial development policy. In the German language this may be partly related to the following reasons: On the one hand "territorial" has a linguistic parallel to the French general and specifically administrative-legal language. On the other hand *'territorial'* e.g., in the context of (GDR-) "Territorialplanung", has always been terminologically avoided in documents in the Federal Republic of Germany, in particular because the German language has a proper linguistic equivalent ("Raum", "räumlich", "Raumordnung", etc.).[22] In

[19] TSPDoc (2006) p 71

[20] A respective regulation has become part of Art. 36 European Charta of Basic Rights, and as such has been transferred to Art. II 36 DraftTEC

[21] see Doc. Com/2000/0580 final

[22] see David (2005a) pp 1151ff

any case, these general economic services are very close to the spatial planning policy aspects of providing equal conditions in spatial infrastructure (traffic, communication, education, etc.). Otherwise it is not easy to determine how closely Art.16 TEC is connected to the wide range of cohesion competences regulated in Title XVII (Economic and Social Cohesion). This title, comprising the Art.158-173 TEC, primarily forms the legal basis for the regional/cohesion policies within the realm of DG XV. Thus, under the current TEC legislation, TC is not an explicit part of social and economic cohesion. Nevertheless the DG XVI tries to include TC in its political activities, but it ostensibly does not want to provoke conflicts within the proper realms of other DGs or of the member states. An advantage for co-ordination efforts is that the addressees of the co-ordination are at the same time probable beneficiaries of the (regional/cohesion/urban policy) funds, which are distributed under the responsibility of DG XVI.

Territorial Cohesion in the Draft of a Treaty Establishing a Constitution for Europe (DraftTCE)

The DraftTCE transfers the regulation of Art. 16 TEC to Art. II 36 DraftTCE[23], but expands TC further to the pivotal regulations respective to the goals and distribution of competence:

- Determination of additional constitutional goals: to promote 'territorial cohesion' (Art. 1, par. 3 DraftTCE: "it shall promote economic, social and territorial cohesion [..]")
- Determination of an explicit (shared) competence in the 'principal area' of "economic, social and territorial cohesion" (Art. I 13, par. 2, Art. III 116 DraftTCE)

The regulations of Art. III 116 – 120 DraftTCE which would replace the current Art. 158 – 162 TEC consequently include TC as part of a threefold cohesion policy explicitly covering economic, social and territorial cohesion. In the view of the Commission, the new Constitution would have transferred "a more stable and less disputable legal base for its political and administrative work" to DG XVI. At any rate, in respect to both the other DGs and to the member states, the legal situation under the new Constitution would not be substantially changed.

[23] Art. II 36 DraftTCE: "The Union recognises and respects access to services of general economic interest [..] in order to promote the social and territorial cohesion of the Union."

2.4 Main Facets of the TC-Issue

The need for co-operation, given the undetermined distribution of power and competence

Territorial Cohesion is related to a set of objectives and goals on various levels of abstraction, which need to be specified and implemented by both co-ordinating (cross-sectoral) and sectoral instruments in the competence of a wide range of public and private-sector actors, which in official policy documents are sometimes called *stakeholders*.

The role of the EU Commission, in particular DG XVI, and the EU Ministers who are responsible for regional/spatial development, is specific to the issue of TC as both have related powers and competences in the field of spatial development. A wide range of powers and competences undisputedly remains with the member states;[24] but significant powers and competences with regard to budgetary funding remain in the EU. There is, however, a disputed, overlapping area of power and competences, which, for various reasons, remains explicitly undetermined in the mutual interest of all concerned.

'Comitology' earmarks a fairly new general strategy of the Commission; the argument being that harmonising EU internal organisational and procedural structures would simplify and make more transparent the existing system of various different types of participation/co-operation. The TC debate was consequently linked with these new organisational structures and brought this new comitology philosophy[25] into line, in order to attribute a better managing control to the Commission.[26] Member states, however, followed the new line and took up their collaboration in the new committees and kept their intergovernmental independence by continuing the post-ESDP-debate in an organisational structure attributed to an 'Informal EU Council of Ministers for Spatial/Regional Planning'.[27] The Ministers continued their work, in spite of a refusal of financial and bureaucratic support by

---

[24] This is why conferences of these EU Ministers are designated as meetings (sometimes as 'informal meetings') of the *'Informal European Council of Ministers responsible for Regional/Spatial Planning'*, as regional/spatial planning is not considered a proper EU competence, substantially belonging to the competence of member states. The designations are rather ambiguous as are the translations into EU-membership-languages. See further 'Resp. to Ambiguity', Arnim v. (2006) Das Europa Komplott, Wien, p 48, referring to the multitude of organisational forms of activity under the designation of the European Council.

[25] TSCdoc (part. C, chap.5, subs.7) p 21. The present version of the document is discussed by the CPG (Coming Presidents Group), and by the Informal Conference of Ministers for Spatial/Regional Planning (supported by an editorial unit). The CPG represents in particular the member states in charge of past, present and coming EU presidencies. It co-ordinates the content, and prepares graphics and maps etc.

[26] Sub-Committee on Spatial Development-CDRC SUD. SUD is the author of the above-mentioned (see footnote 7) expert document

[27] see footnote 24

the Commission. The editorial work on the aforementioned TSCDoc is a substantial result of this continuity, but is currently, as mentioned, still in a draft status.

For both the EU and the member states TC is not at all an isolated policy aspect. It is a co-ordinating and controlling policy within the context of other related legal and political principles, goals, objectives and procedural/organisational measures. Thus far, TC must be understood in a dialectic way, where clearness and ambiguity are mixed. Scientific research and terminology cannot determine a strict borderline between sectoral and cross-sectoral aspects. Approaches to determining whether European Development Policy should be interpreted according to a German, French or a third, perhaps Dutch, conceptual understanding[28] cannot distinctly be related to one member state's philosophy, because European Development Policy comprises elements of all of them. It has various elements with a changing structure over the course of time. It is in the member states' interest to avoid too much regulation, and to keep the TC a flexible instrument.

Non-monolithic structure of the key actors

In particular, it is the financial tools of the EU coherence policy, with substantial EU budget resources distributed through a sophisticated EU funding system, which changes TC from an idealistic concept into a policy with a rather strong political impact. In this respect TC is connected to funds collection and distribution, in terms of the organisation, procedures and policy (e.g., the basic principles to be applied) on all legislative and governmental levels (for the budgetary context see below 2.4; for the governance context see chapter 2.5).

Neither the EU nor its member states should be considered as monolithic bodies. Inside of their governmental structures various bodies/organs/administrational entities equipped with a considerable amount of independence due to their constitutional status (according to the EU/EC treaties) are involved in the legislative and administrative process. They used to have controversial positions due to their political power-conserving interests. On national and supra-national levels this needs to be balanced before agreements can be reached, or it can even lead to failure and in most cases to *'package solutions'* which sometimes reflect political compromise rather than rational considerations.

Budgetary context

Thus, the issue of TC was substantially connected to a compromise for the proportion/percentage of the member states' budgets to be contributed to the EU-budget, forming the financial base for feeding the EU regional/cohesion funding system. In member states such as France and Germany, a restriction on annual national contributions to the EU budget under the inner-governmental responsibility of the respective Ministries of Finance had important effects on member states' spatial policies, by overriding and controlling the policies of the respective ministers in

[28] David (2004) pp 146ff; AIDRU (2005) pp 181-194

charge of spatial planning (France: "Aménagement du Territoire"; Germany: "Raumordnungsminister").[29] On the other hand, the Ministers of Finance's political interest in spatial planning/development boosted its usually weak political position, in particular in its function as a cross-sectoral co-ordination tool.

The inner-governmental responsibilities for spatial planning should be identical with the responsibility for urban and regional planning. This varies from member state to member state, and is dependent on the inner-governmental distribution of responsibilities between the various ministerial departments. Thus, in Germany, urban affairs are together with spatial planning ("Raumordnung") in a ministry bundling responsibilities for transport, construction and urban affairs[30]. Regional economic planning ("regionale Wirtschaftspolitik"), however, is administered by a different department[31] and, of course, the substantial budget responsibilities are under the control of the Ministry of Finance. The fact that the Foreign Office is in charge of EU representation makes it even more obvious that the internal decision-making process concerning spatial planning issues by member states is loaded by inner-governmental conflicts and controversies. In most cases there is no public notice or information.

2.5 Governance Aspects of Territorial Cohesion

In the TSPDoc[32] the governance issue is addressed in the broad definition of the scope of TC as 'governance philosophy', describing the concept of how TC could be implemented within the framework of 'territorial governance'. This refers in particular to the distribution of rules and responsibilities among the different levels of government (supra-national, national and sub-national) and the underlying processes of negotiation and consensus building.[33] In addition to this, however, the internal structures of EU and member states' institutional bodies and sub-bodies are part of the governance issue.

The diagram shows the ambition of EU cohesion policy to be extended to multi-level governance (EU, national, regional, local level); in particular involving the relevant sector policies (transport, agriculture, R&D, environment, etc.). This strategic approach is very similar to the environmental policy approach, but differs insofar, as that up to now territorial cohesion has not had the explicit powers legally requiring cross-sectoral consideration that environmental policy does.

---

[29] David C H (2005)

[30] "Bundesministerium für Verkehr, Bauen und Wohnen" (Federal Ministry of Transport, Building and Urban Affairs). Spatial planning appears neither in the German nor in the English designation of the ministry and has a sub-departemental status in Germany.

[31] Bundesministerium für Wirtschaft (Federal Ministry of Economics).

[32] TSPDoc 2005/2006, Part A: Defining the Scope, 1. Territorial cohesion and the added value of territorial development policies – governance policy.

[33] OECD (2001) p 135 and p 142

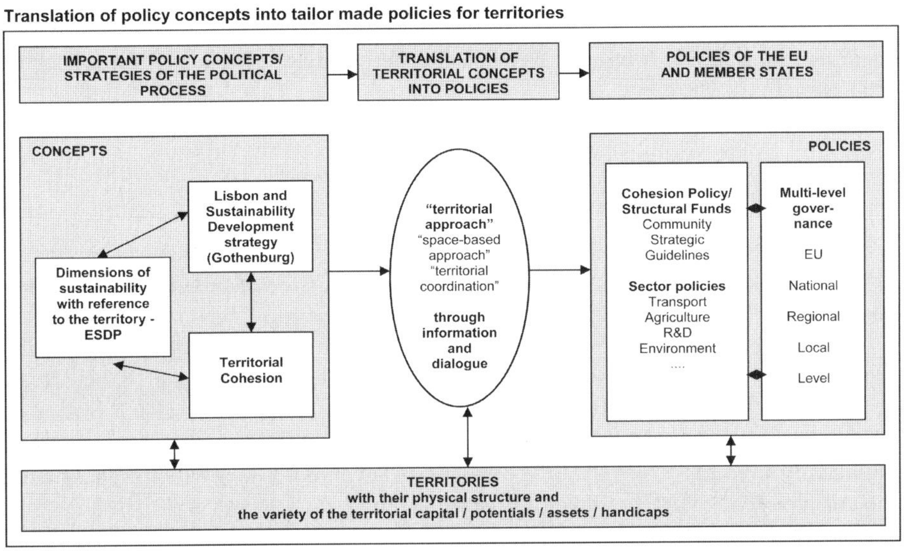

Fig. 1: Diagram taken from the TSPDoc; in TSPDoc 2006 it is not included anymore. Source: TSPDoc (2005) p 11

The Commission, in particular DG XVI, expects that if a formal shared competence of the Union and member states would be created through the ratification of the EU Constitution, this would strengthen the European mandate for TC, thus backing the responsibility to promote a coherent approach to territorial development by both the EU and national policies. The governance philosophy behind this is that the Constitution would not substantially change the situation of the allocation of competencies, but might allow the Commission to interpret the explicit shared responsibility as a mandate for more structured EU (i.e., Commission-run) initiatives.[34] The Commission's current TC argumentation is under strong political pressure caused by the constitutional debate, with regard to the application of the principle of subsidiarity.[35] Thus, the Commission avoids initiatives which could clearly be interpreted as a political approach to expanding EU responsibilities to the disadvantage of the existing member states' responsibilities. The Commission therefore restricts itself to inner (EU) organisational approaches, which are without doubt within its realm. Although EU internal ('inter-service') co-ordination of the various EU policies with territorial impact is explicitly addressed as an object of TC policy, DG XVI has a competence in one of these sector policies: regional/

[34] TSPDoc (2005) p 13

[35] See the protocol on the application of the principles of subsidiarity and proportionality amending the text of the Constitution, derivative of EU 2003/ C 169/95, 18.7.2003; see further: David (2005a) pp 1129-1134

coherence policy. At present the Commission avoids claiming a comparably strong co-ordination/control role as it has in other cross-sector related policies, such as environmental policy.

Instead of this, it resorts to the future EU cohesion policy strand of European Territorial Co-operation (abbreviated in theTSPDoc as ETC in this context). The reason given for this is "that its focus, the trans-European dimension, plays a key role both in strengthening the structure of the European territory and in promoting better territorial governance in the EU."[36] Asking only for co-ordination ostensibly avoids any harder internal collision within the Commission and appeals to voluntary co-operation between the DGs involved.

The extension of this governance philosophy to all levels of administration down to the local level is another questionable issue. Considering the TC objectives respective to the Lisbon and Gothenburg ambitions, from the supra-national (i.e., EU) and national levels implicates the creation of a rigidly restricted funding policy which only supports projects in line with this and excludes issues which do not fit into the concept. Thus the local level, in particular (e.g., peripheral) areas which expect more equal infrastructural endowments from TC, will often face disappointment if other better-off areas receive funding. The lessons of good-practice projects cannot be transferred everywhere. The coming stakeholder conferences in 2006 and 2007 will probably make the political expectations connected with TC on local and regional levels more transparent. Some participants will probably face the prospect of falling through the grid created in the Lisbon and the Gothenburg strategy.

The debate about the distribution of regional/cohesion fund resources in terms of the percentage that can be expected for member state territory overall, and for certain designated areas in particular, will certainly be revived on the more abstract level of TC policy, even if the existence of a direct link between TC and fund distribution is officially and explicitly denied.

TSPDoc argues prudently that TC might even apply to areas of weak economic structure or to those physically or geographically handicapped.[37] This may, however, be of little comfort to those regions and localities not optimally qualified when watching other better-off areas receive support.

Negative structural consequences will be experienced on a national policy level rather than by the EU as a whole, which will appear as a benefactor to those few fortunate enough to receive EU funds. The responsibility for those who receive no EU funds, however, remains with the member states. German constitutional experience shows that Federal State ("Länder") co-operation supported by selected federal financial aid tended to corrupt the federal system, so that the federal system with its clear distribution of (in particular financial) competences and powers then had to be restored. This is currently again an actual subject of a reform of the

[36] TSCDoc (2005) p 21 (6)

[37] So does Michel Delebarre, President of the Committee of the Region, in his address to the Amsterdam shareholder-meeting on 28 June 2006, *The Lisbon Strategy and TC: Towards a New Kind of European Governance.*

German federal system ("Föderalismusreform"). It shows that ideas of co-operation cannot replace the need for a clear delimitation of the range of powers in case of overlapping areas of competence.

2.6 Achieving Territorial Cohesion through the Results of Scientific Research

Scientific co-operation has proved to be a possible and reliable way to reach a mutual understanding of a political situation and to come to a respective resolution during the ESDP preparation procedure. This was one of the reasons behind the establishment of the ESPON programme, which is not dealt with in this paper.

As the preceding diagram shows, this strategic idea of implementing TC has been chosen under the designation "territorial approach through information and dialogue." In recent political debates key words appear, such as 'evidence-based policy' or the use of 'evidence-based synthesis documents'.[38] The series of stakeholder conferences which will occur before the envisaged final TC policy paper is adopted during the German presidency in 2007 are intended to fit into this strategy.

From a scientific point of view, the mixture of policy objectives and scientific methods or findings raises some questions. Determining political goals and objectives can and should be rationally orientated to scientific findings. But political directives cannot be substituted through research findings.

The elaboration of the 3rd Report on Economic and Social Cohesion in 2004 was substantially based on ESPON project findings, but is independent from these findings in its political conclusions and propositions. The ESPON researchers were not involved in the elaboration of the report, nor was it made transparent in what respects the report follows or diverts from the research base provided by ESPON. The selection for the report was made politically, and not scientifically. This is not an allegation; it merely states how research-based policy is used. The borderline between policy and research should be kept in mind, however.

2.7 Normative and Governance Perspectives of TC

Linking TC with the Lisbon and Gothenburg considerations implies that the political fate of TC is dependent on long range political objectives. According to a widespread political assessment, these objectives may not be reached in the near future or may never be reached at all. The authors of the TSP-document were wise enough not to speak of Lisbon and Gothenburg goals or objectives, but merely of

---

[38] The designations appear already in the resolutions of the informal ministerial meeting in Luxemburg, 20/21 May 2005

'ambitions'. This may be a reflection of their well-founded strategic scepticism.

The political attractiveness of TC for many European regional and local authorities (and their citizens) arises less from long-term political expectations, but instead from the attractiveness of receiving financial funding for short or middle-term projects which are partly or fully financed through EU programs in the realm of various General Directorates. The interconnection between the underlying funding system and TC is in many cases not very close.

Sometimes it may only be affected through DG XVI's attempted political approach of establishing TC as a co-ordinating policy objective which would allow them to exercise control functions respective to the other territory-related national and EU policies. This implies an extension of the proper sector competence (for Regional/Coherence Policy) to a cross-sectoral competence (as the German "Raumordnung" defines itself), which is in its legal dimension more disputed than distinct. In the past, the EU's internal organisational DG structure has been fairly resistant to any EU-internal co-ordination, and even if agreement is reached on the common Lisbon/Gothenburg considerations, this will presumably bring about no change. Scepticism applies thus far.

The EC funding of specific economic projects (this can be applied to many research projects too) based on procedural mutual involvement of the Commission and selected regional and local bodies ('stakeholders') will continue to intensify the bypassing of standard channels of co-operation between the Commission and member states, as regulated in the EU/EC treaties. This tendency will be politically supported by the European Parliament and by other EC bodies, e.g., the Committee of the Region, which until now has not had the legal status of an EC institution, nor have its members any EU law-determined status.[39]

The discrepancy between the specific project-related involvement of the Commission (i.e., DG XVI), allowing punctual political profiling, and the continuity of general responsibility resting with the various levels of member states' governments and administrations (e.g., regional and local development or urban policy) leads to a contradictory picture of TC instead of producing a consistent, publicly understood and agreed-upon policy pattern.

Given this general situation and due to member states′ competition for EU-funding, 'intergovernmental' co-operation within the informal European Council of Ministers for spatial/regional development, which is operating beyond the formalised EU comitology channels and procedures, will result in agreements which can be characterized by a high level of abstraction and a low overall regulatory effect rather than by formulating operational criteria for TC. In fact, those formal regulations directly controlling the distribution of EU (regional/cohesion) funds,[40] which are within the clear competence of the Commission, are in the political focus.

[39] see Doc. EP, Resol, 14 January 2003, 2002/2141, in reference to the so-called *Napolitano Report*, on the role of regional and local authorities in the European integration (2002/141), further, Committee of the Region Doc. 237/2002

[40] see e.g., the respective regulations in the *Official Journal*, L410, vol. 49, 31 July 2006

As long as the relevance of TC for the formulation of these regulations is indefinite or even non-existent, political decisions made by the European Council of Ministers for regional/spatial planning will receive only limited political appreciation. In the medium and long-term the EU comitology system presumely will prevail over intergovernmental co-operation. It must be taken into account that the national ministers represented here not only have to distinguish themselves with regard to the Commission and its respective DGs, but moreover with regard to their national, inner-governmental ministerial colleagues, who are not at all politically willing to consent to cross-departmental policy approaches, which is what TC purports to be.

2.8 Going back to the Roots of TC?

Going back to the conceptual roots of TC, as found in early French publications,[41] ostensibly the idea of TC cannot be fully dismantled from the idea of creating cohesion through efforts to diminish existing (economic, social and territorial) disparities.[42] Disadvantaged regional and local entities must at least receive a long-term perspective, for example, ultimately being attached to the infrastructural status of the better-off regions and cities. In the German constitution this is addressed by the constitutional term 'equivalence of living conditions' ("gleichwertige Lebensbedingungen", Art. 72, par. 2 GG). It must be made clear that this is an ideal goal which might only be reached in the far future or never at all. It does, however, describe the need for a consistent concept covering the totality of the territory in a view contrary to reducing TC to a set of selected localities and projects matching the Lisbon/Gothenburg ideas. Rather than attracting stakeholders' attention by expectations of funding, consent to TC should be pursued by a coherent policy approach transforming TC from a more sector-orientated policy to a cross-sector co-ordinating and controlling competence. Only a TC policy which is transparent to the public and respects both the principle of subsidiarity and the governmental structure of the member states will get permanent political acceptance by the public stakeholders and citizens needed for its implementation.

---

[41] see the basic analysis by Husson, *L'Europe sans Territoire* (2002), summarised by David (2004) above footnote 7 and footnote 22

[42] DG XVI intensifies to use urban policy as a new political approach of cohesion policy to attract a broad public interest; see EU Doc. Cohesion Policy and the Cities, a public consultation project (2006); further the consultation with reference to cohesion policy in support of growth and jobs, Community strategic guidelines 2007-2013, Report (2005)

References

AIDRU (Priet, Univ.Orléans) (eds.) (2005) La formation d'une politique de developpement de l'espace communautaire: convergence des politiques francaise et allemande, mélanges en l'honneur de Henri Jacquot, Presses Universitaires d'Orléans, pp 181-194

David CH (2006) La formation d'une politique de developpement de l'espace communautaire: convergence des politiques francaise et allemande. In: Mélanges en l'honneur de Henri Jacquot, Presses Universitaires d'Oléans, pp 181ff

David CH (2005) Zur Konvergenz der Raumordnungspolitiken Frankreichs und Deutschlands im Post-EUREK-Prozess. In: *Raumforschung und Raumordnung*, Issue 1/2005, pp 11-20

David CH (2005a) "Subsidiaritätsprinzip" and "Territorialitätsprinzip" (keywords). In Handwörterbuch der Raumordnung, 2005, pp 1129-1134 and pp 1151ff

David CH (2004) Territorialer Zusammenhalt: Kompetenzzuwachs für die Raumordnung auf europäischer Ebene oder neues Kompetenzfeld? In: *Die öffentliche Verwaltung* (2004) pp 146ff

David CH (1993) Europäische Tendenzen und gemeinschaftliche Grenzen einer Harmonisierung raumplanungsrechtlicher Vorschriften. In: *Die öffentliche Verwaltung* (1993) pp 1021ff

ECE (Commission of the European Communities) (2005) Community Strategic Guideline for Cohesion 2007-2013, Working Document of DG Regional policy

ECE (2004) 3rd Report on Economic and Social Cohesion, Luxemburg

ECE (2001) White Paper on Governance

European Commision (1999) European Spatial Development Perspective ESDP

European Community (2001) Treaty of the European Community (TEC), finally amended in Nice

European Convent (2003) Draft establishing a Constitution of Europe. Doc. 2003/ C 169/01

European Parliament (2002) Napolitano Report, Resol 14-01-03. Doc . 2002/2141

European Union (2006) *Official Journal*, L 410 2006 resp. EU Regional Fund Regulations

Husson C (2002) L'Europe sans Territoire. Paris

OECD (2001) Territorial Outlook - Territorial Economy, Paris

3 Territorial Cohesion – Between Expectations, Disparities and Contradictions

Maroš Finka

The concept of territorial cohesion extends beyond the notion of economic and social cohesion by both adding to this and reinforcing it. In policy terms, the objective is to help achieve a more balanced development by reducing existing disparities, preventing territorial imbalances and making both sectoral policies, which have a spatial impact and regional policy more coherent. The concern is also to improve territorial integration and encourage cooperation between regions.[1]

Introduction

Current spatial development policy in EU countries has been intensively influenced, in addition to economic and social cohesion, by a new concept – the concept of territorial cohesion (TC). As the term '*territorial cohesion*' was initially introduced in the political sphere, it has evoked, independently from the process of new treaty ratification, many discussions about the substance of this concept. The task of elaborating the concept and translating it into European policies intensively has occupied politicians, practitioners and academicians since the term "territorial cohesion" first appeared in the proposal for a new treaty. EU informal ministerial meetings on territorial cohesion in Rotterdam (2004) and the Informal Meeting of Ministers for Regional Policy and Territorial Cohesion in Luxemburg (2005) focused on the interlinks between the territorial cohesion concept and the Lisbon-Gothenburg strategy and defined the political direction in the European discussion on this topic. The Luxemburg meeting stressed the importance of stimulating a broad interdisciplinary EU dialogue on territorial cohesion with EU institutions and local and regional actors which would reflect the reality of the delaying fundamental interdisciplinary professional discussion, thus creating the theoretical background for territorial cohesion in terms of a political spatial development concept.

Although the ongoing ESPON Programme[2] has been an achievement in terms of data collecting, and to a certain extent of vision making, it cannot create a satisfactory professional background for the formulation of the contents of a new political concept without putting the results of the programme forward and linking

[1] Laissy (2004)

[2] ESPON – The European Spatial Planning Observation Network was set up to support policy development and to build a European scientific community in the field of territorial development, see www.espon.lu.

them to a certain theoretical concept. It will therefore be of great importance to undertake further scientific work with the aim of drawing coherent conclusions from the data and making recommendations under the light of conceptual integrative viewpoints.

3.1 The Spatial Development Process and Territorial Cohesion Concept

The spatial system of the European Union is a cultural phenomenon created by the interaction of nature and society over the course of a thousand years. The spatial development process is a process that reflects the development of society as well as the development of its needs concerning:

- The appropriate spatial organisation of society
- Specific features of spatial structures in the physical environment for social processes
- Conditions for the interaction between society and nature with the goal of optimising social processes, of which production and distribution play a leading role

The development of European society[3] at the end of the 20th and beginning of the 21st century brought new imbalances and tensions which resulted from the contradictions between the different dynamics of social development and the development of spatial systems. The probability of the appearance of these imbalances grows with the dimensions and dynamics of the system. Dynamic processes of societal transformation towards a knowledge-based information society, the raising of ecologic awareness, European integration, EU enlargement and world-wide globalisation opened a gap between the changed needs of society and the inertia of the natural and man-made structures of the physical environment.[4] The complexity and the intensity of internal interrelations increased the danger of an emergence of instability in the system as a whole, caused in particular by imbalances.

The response of EU policy and national policies of EU member states to the existing problems covers a broad scale of options in various fields. Most of them have been immediately oriented to the minimalisation or elimination of the imbalances perceived by the public, often without an awareness of the spatial/territorial aspects of the intervention, or without recognition of the effects to the main stream caused by the adaptation of spatial structures to the current needs of the society undergoing transformation towards a knowledge-based information society (agricultural policy,

[3] For example integration processes, EU enlargement, transformation processes towards the creation of a knowledge-based information society, etc.

[4] Finka/Petrikova (2002)

transport policy, labour-market protection policy, etc.). The effects of many of these policies were inefficient, and in contradiction with comprehensive cross-sectoral goals or with particular goals in other fields of policy.

The tension between new development impulses, which occured upon EU enlargement, and the negative effects of forthcoming globalisation, shortages of natural resources, a decline in the competitiveness of the European economic space, the appearance of new development poles in the world, and an increased danger of terrorism are only some of the factors catalysing an intensive EU discussion on the new environment for the development of a knowledge-based information society in Europe. This discussion, which generally revolves around three sectors – economic, social and institutional, was intensively influenced by the historical model of integration in the EU (*'integration sui generis'*), with a lack of consensus or even discussion about the final quality targeted by the integration process[5], which is closely connected to questions concerning the quality of spatial/territorial structure within the EU. The achievement of a new spatial development quality in the EU, reflecting the needs of a sustainable development of a knowledge-based society in an enlarged EU, is closely connected to searching for a balance between territorial sovereignty and optimisation at the supranational level. In this context, it might be surprising that the term *'territorial cohesion'* was included in the proposal for the EU Constitution, although this can be understood as a definition of one aspect of the new required quality of spatial structures in the EU, respecting the needs of European society, which is being transformed into a knowledge-based society as defined in the Lisbon Strategy.

The introduction to the concept of territorial cohesion highlights the need to take into account current problems resulting from the above-mentioned development as well as the diversity of the European continent, in order to achieve the rich and complex meaning of cohesion. Each territory has its own assets and faces constraints that development policies have to take into account to be efficient. In this context, there is the need for a broader view of cohesion, encompassing all dimensions of spatial development and horizontal interactions. The Galway Conference on territorial cohesion (25-27 May 2004) defined territorial cohesion thus: *"As a policy objective, territorial cohesion aims at contributing to the harmonious and balanced development of the Union by reducing economic and social disparities, by preventing territorial imbalances from emerging, and by making sectoral policies that have a spatial impact more coherent due to regional policy. Territorial cohesion also aims at improving territorial integration and encouraging co-operation between regions."*

This interpretation of the term *'territorial cohesion'* has very little to do with the quality (in an axiological sense) of the territorial/spatial systems. The territory only seems to create a reference structure for measuring the economic or social indicators, or for balancing inputs and outputs. Even the integration of co-operation, in addition

[5] Gustedt (2005)

to *'cohesion'*, *'coherence'* and *'co-ordination'*, into the quality of *'territorial cohesion'* does not create the potential for creating a new quality, as the added value of the implementation of the territorial cohesion as a new objective in the policy practice. The concept of territorial cohesion policy, as it has been presented so far[6], appears not to represent any principally new approach. It can be understood more as an expression of a reflected need to follow the system linkages between different dimensions of cohesion in the territorial socio-cultural systems in the EU, and as such does not bring any new contradictions or disparities into the practical policy. Does this really represent everything that can be brought in by introducing this term and concept in the uppermost EU policy for territorial development in the EU?

We have to begin at the assumption that territorial cohesion is not only a political watchword, but that it can be characterized as a political and "theoretical concept incorporating the concern for co-ordination of the comprehensive integrated approach and concern for actively pursuing balanced development throughout the territory concerned like in the regional economic approach"[7].

For spatial planning, as a scientific discipline and planning activity included in spatial development management, the new quality of this theoretical concept expressed by the term *'territorial cohesion'* has to follow three principal dimensions:

(A) Quality of the territorial structure as defined by:

The optimal utilised territorial capital and development potentials of territorial units

Concerning this dimension, territorial cohesion is associated with 'opening-up the territory', identifying and strengthening the economic development potential of all territories in order to activate their endogenous potentials for achieving sustainable economic growth. Location-based comparative advantages, or territorial capitals, are important assets for endogenous development. Thus, to achieve a high level of global competitiveness, the utilisation of these territorial capitals can be regarded as one of the central aims connected to territorial cohesion.

The optimal organisation of territorial structure

The reference quality for optimising the organisation of territorial structure is the quality of conditions which the territorial structure creates for social processes. The development of a knowledge-based information society, defined as the prospect and aim of EU development in the Lisbon Strategy, introduced new requirements concerning the quality of territorial structures, their organisation, and the quality of elements. In addition to the often-discussed spatial innovations like new space-

[6] for example, in: Hauptausschuss der Ministerkonferenz für Raumordnung (MKRO) (2005); Bundesministerium für Verkehr, Bau und Wohnungswesen (2005); Office for Official Publications of the European Communities (2004)

[7] Faludi (2004)

time structures, virtual spatial structures, self-learning structures, and intelligent cities or regions, one of the challenges concerns territorial cohesion directly. It is the requirement to create the optimal environment for the development, transfer, and implementation of know-how in the EU – for the creation and dispersal of innovations. Despite the diversity of innovations, they have at least one common requirement concerning an environment optimal for their distribution – the quality of the linkages between the elements of the territorial system (i.e. density, intensity, capacity), which are the features of the organisation of the system integrated in the quality of *'territorial cohesion'*.

A crucial question in this context is, which type of territorial organisation allows the achievement of appropriate conditions for the development and spread of know-how, which type of territorial organisation allows for the achievement of territorial cohesion. A precondition for the activation of territorial capital, as the basis for endogenous development, to respect diversity, as one of territorial features and preconditions for sustainability. The profiling, based on the diversity of the territorial capital and co-operative competition, should be the basis for the organisation of the territorial structure at the regional and supra-regional levels. The organisation of the system has to create conditions for optimising the use of the endogenous potential of localities (urban, rural), settlement units (cities and villages), and regions (rural, urban, metropolitan, etc.) in the supra-regional units – the national territorial systems in the EU and in Europe. In this context, territorial cohesion cannot be exclusively bound at the EU regional policy level but must have broader spatial implications. EU territory is not isolated and the organisation of its territorial structure cannot ignore the development beyond EU borders, and certainly cannot ignore the further development of the EU itself.

Territorial cohesion is an issue which not only concerns the EU's internally oriented policies, but also concerns cohesion of EU territories with neighbouring territories outside the EU (the opposite of the ESDP).

The quality of equal living conditions on a territorially balanced basis
In our opinion, the notion of territorial cohesion combines integration and solidarity (equity), which corresponds to the concept of spatial justice developed by Jaques[8], thereby extending and applying Rawls' (1971) principle to territorial development. It could therefore be defined as one of the policy objectives governing EU policies as a whole. Territorial cohesion as the new quality of European spatial structures does not mean less disparities and more homogeneity. Each territory can play a specific role in the system as whole, based on its specific features and uniqueness. Cohesion in the territorial system means that:

[8] Lévy (2003)

- The organisation of the system (spatial organisation included) provides optimal conditions for the existence of the territorial units in the system (e.g., of the municipalities in a region)
- The interaction of territorial units in the system[9] creates optimal conditions for the capitalisation of their territorial potential
- Each territorial unit has equal access to the possibilities and profits produced by the system as a whole thanks to their active coexistence in the system and the use of their territorial capital

This is the foundation for the creation of equal living conditions on a territorially balanced basis. One of the crucial questions in this context is the hierarchical level of territorial units, which can be appointed as appropriate for territorial balancing. The core of the answer lies in the spatial accessibility (offer/supply) and spatial range (mobility) of the inhabitants' activities. The simplification of automatically using the level of the regions (even characterised by the area and number of inhabitants) for the demonstration of territorial cohesion issues without defining what *'region'* means is misleading and has caused inefficient intervention into regional development in the past. As the traditional view basically relates territorial cohesion to territorially demonstrated socio-economic differences and inequalities, a first-level connotation relates territorial cohesion to geographical features, acting as the disadvantages of territorial units. For geographically handicapped territories, like islands, mountain areas, etc.[10] deficits relate to their connection to and interaction with other European territories. A policy in favour of territorial cohesion would have to reduce the negative effects of the geographical conditions and create the possibilities for the capitalisation of them as the specific development potential and potential for the specific functioning of territorial units in the system as whole.

(B) Quality of the processes (functioning) in the territorial systems

The second aspect, which is related to the previous one, refers to growth and decline, competition and co-operation, territorial integration and territorial sovereignty, coordination, subordination and subsidiarity, communication, and the exchange and flows amongst and between territories. The connection of the dimension of territorial structures with the process dimension of the existence of territorial settlement systems seems to be crucial to the definition of the new quality, represented by territorial cohesion. On one hand the focus of social cohesion is strongly directed at the individual level of persons or households, to the minimizing of the poverty and unemployment of individuals. Economic cohesion is, on the other hand, oriented towards the intermediate level of institutions (enterprises, unions, etc.) and institutional settings (e.g., tax systems). Territorial cohesion represents territorial context and territorial aggregates (predominantly regional) of the quality 'social and

[9] Territorial units fulfilling their more or less specific function in the system.

[10] cf. Third Cohesion Report

economic cohesion' referring to the territory as a frame for social processes and an aggregate of individual situations. Territory as the holder of this new quality creates the frames for social processes, creates the existential context for individual being and acting, and is the reference phenomenon for the integration of individual acting into social processes.

The concept of territorial cohesion, as it is characterised in the official documents, completely exposes the main contradiction of the EU policy in the field of spatial development. On one hand there is the EU's ambition to intervene and work towards spatial development optimisation, supported by a set of very strong tools (cohesion funds, structural funds, etc.), in partial sectoral policies (transport, ICT, social infrastructure, environment, agriculture, regional development, etc.). On the other hand there is the EC's lack of responsibility for spatial development, expressed in formulations concerning territorial cohesion which stress the Open Method of Co-ordination[11] as the main instrument.

The integration of the structural and process dimension (the quality of structure, structural elements and the quality of their interactions and development processes) is the key issue for safeguarding the sustainability of spatial development in the EU's knowledge-based information society.

To understand the interlinks between the processes in society and the (required) quality of territorial structures is the precondition for the right timing and efficiency of immediate interventions into spatial development and also into societal processes. The current policy of agricultural subsidies as part of the management process at the EU level is an example of these unhealthy interventions, which deform the conditions for an optimal use of the endogenous potential of territorial units in the EU, and even cause environmental damage. In this context these kinds of sectoral interventions are counterproductive in relation to sustainable territorial development in the EU.

Competition and co-operation of territorial units are two inseparable categories and are necessary for the cohesion of the territorial system. The sense of co-operation can be seen in the improvement of the potential of co-operating subjects to efficiently fulfil their roles/tasks and provide the necessities of life for citizens in a sustainable way. The extent to which they fulfil these roles/tasks identifies their attractiveness and is a precondition for competitiveness. An improvement in attractiveness is one of

[11] Medium term between the Community and intergovernmental methods, created as part of employment policy in the Luxembourg Process, defined as an instrument of the Lisbon Strategy. The OMC provided a framework for co-operation between the member states, whose national policies can thus be directed towards certain common objectives. Under this intergovernmental method, member states are evaluated by one another (peer pressure) with the Commission's role being limited to surveillance. See also: 'Lisbon Action Plan Incorporating the EU Lisbon Programme and Recommendations for Actions to Member States for Inclusion in their National Lisbon Programmes'. Companion Documents to the Communication to the Spring European Council 2005 / Working Together for Growth and Jobs, COM(2005) 24 final (EC Brussels 2005).

the main spheres of competition and concerns each territorial unit and as well as the EU as a whole. Territorial cohesion as the new quality refers to the potential of EU territories and their improvement, their activation and the mediation/improvement of accessibility for internal and external demand holders.[12]

(C) Territorial subjects of the societal development process

Challenges resulting from this new quality defined as *'territorial cohesion'* refer to different territorial subjects (municipalities, self-governmental regional units, states, the EU) or territorially bound subjects (inhabitants, enterprises, politicians, administrative authorities and other stakeholders). It seems to be very clear that achieving the aims defined by the political concept of territorial cohesion up to now[13] should not be based on fulfilling the centralised competence or on comprehensive European strategic planning documents. The only possibility is that the concept be based on a continuous process of negotiations between the European Commission, member states, and regions, which might include representatives of sector policies and other territorially acting stakeholders, and is based on a commonly shared reference system in terms of goals, options, concepts, and strategies for territorial development.[14]

Although the Commission has developed the concept of 'multilevel governance' in its White Paper on Governance, which proposes to exceed the narrow design of subsidiarity and seeks the means of a Europe/States/Regions coordination, the concept of territorial cohesion requires clearer articulation between Europe, national states, regions and communities, and more explicit coordination between sectoral policies at all levels.

3.2 Territorial Cohesion, EU Objectives and current Development Goals

The concept of territorial cohesion has been positioned by politics as a tool supporting the achievement of EU objectives and current development goals. The reference base for the assessment of the territorial cohesion concept has to be, in this context, the European Union's set of main strategic development goals and superordinate objectives – equity, competitiveness, sustainability, and good governance.

[12] Finka (2005)

[13] see, for example, "... In policy terms, the objective is to help achieve a more balanced development by reducing existing disparities, preventing territorial imbalances and making both sectoral policies which have a spatial impact and regional policy more coherent. The concern is also to improve territorial integration and encourage cooperation between regions." (Ministerial Meeting on Territorial Cohesion on 29 November 2004 in Rotterdam, press report, Ana-Paula Laissy).

[14] Sozietät/Nörr/Stiefenhofer /Lutz/Battis (2005)

The main strategic goals for the current phase of EU development are defined in the Lisbon Strategy, in the document "Employment, Economic Reforms and Social Cohesion – Towards a Europe Based on Innovation and Knowledge" (2000) and based on a mid-term review of the implementation of the Lisbon Strategy, specified in documents prepared for the Gothenburg EU council meeting "Partnership for Growth and Jobs" (2005), and supported by the "Union Action Programme" and "National Action Programmes".[15]

Based on critical assessment of the implementation process of the Lisbon Strategy, the European Commission proposed a new start for the Lisbon Strategy, focusing efforts around two principal tasks – delivering stronger, lasting growth and creating more and better jobs. Meeting Europe's growth and jobs challenge is the key to unlocking the resources needed to meet the wider economic, social and environmental ambitions; meeting those wider goals will anchor the success of the reforms. Three fields of priorities were set:

A more attractive place to invest and work
- Extend and deepen the internal market
- Improve European and national regulation
- Ensure open and competitive markets inside and outside Europe
- Expand and improve European infrastructure

Knowledge and innovation for growth
- Increase and improve investment in research and development
- Facilitate innovation, the uptake of ICT, and the sustainable use of resources
- Contribute to a strong European industrial base

Creating more and better jobs
- Attract more people into employment and modernise social protection systems
- Improve the adaptability of workers and enterprises and the flexibility of labour markets
- Invest more in human capital through better education and skills

In relation to the above-mentioned reference basis, the strengths concerning the territorial cohesion concept, as it has been presented so far, can be highlighted. The concept of territorial cohesion has ambitions to translate the fundamental goals of balanced and sustainable development into territorial settings (integration, building on the ESDP, cooperation aimed at the integration of the territorial dimension into EU policies, taking account of regional diversity, and the challenges of multi-level and multi-sectoral governance). In the ESDP the EU member states have also identified methods for dealing with this territorial complexity.

---

[15] *"The concept of TC can potentially contribute to sustainable and balanced development [..] it translates the goal of sustainable and balanced development assigned by the Union into territorial terms."* Articles 2 and 16 of the Treaty proposal (Nice version).

The first principle is that of horizontal cooperation (between policies and between public and private players) at each level; for example, no one in Europe doubts that public regulation is necessary at the local level for spatial planning; a second principle, less obvious, is that of vertical cooperation between various levels. In applying the principle of subsidiarity, the management of a given territory is the concern of one political entity, and that entity alone. The free movement of persons, goods and capital, however, and the interdependence of levels, which is accelerating due to globalisation, mean that no territorial level can deal on its own with all the issues related to space, and that spatial regulation is becoming a problem shared between several levels. The aim of territorial cohesion presupposes the establishment of co-operation in both horizontal terms (between policies) and vertical terms (between actors/stakeholders at different geographical/administrative levels). It should become a general concern, integrating the territorial dimension into the design and implementation of Community and national policies.

Attempts to introduce the new concept of territorial cohesion shows that European countries share a view in which territories (states, regions, cities) are factors of production (places of identity and of collective purpose) and of solidarity, which is the basis for the existence of different political levels and territorially differentiated policies. In this sense, the notion of territorial cohesion combines coherence (*territorial integration*) and solidarity (*territorial equity*).[16]

The introduction of the concept of territorial cohesion is accompanied by analytical work and a definition of the tasks for monitoring spatial development[17] and providing an "insight into the territorial state of the Union and a common information base to address key challenges and opportunities." The analytical work is followed by proposals for European Territorial Cooperation (facilitating development, focused on facilitating the Lisbon/Gothenburg Strategy, and supporting efficiency in achieving cohesion), stressing the need for a flexible strategic approach incorporating different facets of territorial cooperation.[18]

The territorial cohesion concept is problem-oriented and tries to avoid the introduction of additional bureaucratic procedures or administration (no new procedures or rules, the use of existing possibilities, including EU expert committees, impact assessments, existing council structures and working groups). It is expected that the concept of territorial cohesion will meet with a high level of acceptance by all member states with respect to the political reality of the EU-wide position of territorial sovereignty and the current distribution of competences.

Problems in this phase of development of the territorial cohesion concept can be considered from several aspects. The territorial cohesion concept as it has been presented so far is characterised by a high level of vagueness. There is no clear

---

[16] Peyrony (2005)

[17] see for example: Office for Official Publications of the European Communities (2004)

[18] see for example: European Community (2006)

idea or, to be more accurate, a common understanding of what territorial cohesion might or should mean. Controversies concerning the legal and political meaning of the term *'territorial cohesion'*, and those concerning the effects and side effects to be expected, do exist.

The concept of territorial cohesion as it has been presented thus far does not fulfil the ambition of translating fundamental goals of balanced and sustainable development into territorial settings. The term territorial is sometimes connected with the term a) cohesion – as in the Treaty – and sometimes with b) coherence or coordination (in the sense of compatible, non-contradictory sector policies with regard to their respective intended and unintended impacts), or c) with the term co-operation, as one out of three new objectives of the Structural and Cohesion Funds.

Experience gained from implementing the Lisbon Strategy indicates the problems of sharing the negotiated priorities among EU member states, which do not want the chance of the territorial cohesion concept unless there are strong tools for its realisation.

The territorial cohesion concept merely extends existing policies, but, with a sharper focus on development opportunities, encourages co-operation and networking and with more attention to the strengths of the areas and better targeting of policy instruments. Territory in the concept of territorial cohesion as it has been presented so far seems to create only a reference structure for measuring economic or social indicators, or balancing inputs and outputs. Even the integration of co-operation, in addition to cohesion, coherence and co-ordination, into the quality of territorial cohesion does not offer the potential to create new quality through the added value of the implementation of territorial cohesion as a new objective in policy practice.

The concept of territorial cohesion, as it is characterised in official documents (see above), completely exposes one of the main contradictions of EU policy with regard to spatial development: on one hand, the ambition of the EU to intervene and promote spatial development optimisation supported by a set of very strong tools (cohesion funds, structural funds, etc.), partially through sectoral policies (transport, ICT, social infrastructure, environment, agriculture, regional development, etc.) is evident. On the other hand, the EC's official assertion that it has no responsibility for spatial development is repeatedly expressed in several statements revealed, for instance, in formulations about the so-called Open Method of Co-ordination as a main instrument for the implementation of TC. This method is still under discussion concerning its governing capacity and capacity of harmonization.

The concept of territorial cohesion as it has been presented so far offers no adequate solution in cases where the opportunities or problems are transnational, and require an intervention from the upper level. This is, in the language of economists, where the discrepancy between the administrative frontiers of the states and economic and social reality generates *'externalities'*, which can justify a higher level intervention: European networks (TEN); cross-border cooperation, when the persistence of linguistic, legal or cultural barriers hampers the growth of cross-border conurbations; transnational cooperation implementing networks of cities or development corridors, the sustainable development of transnational sea and river basins or mountain ranges.[19]

The concept of territorial cohesion presented thus far does not appear to be able to support an activation of the unique development potentials of EU regions and does not convincingly fulfil expectations of reducing existing disparities between regions or contradictions in EU policies of spatial/territorial relevance. Attempts using current EU policy tools have been limited as to their effects and the concept of territorial cohesion does not re-assess or expand upon the spectrum of relevant instruments. Of course, any classification of disparities and contradictions depends on their respective interpretation and context.

It is not clear how the concept of territorial cohesion is to contribute to a definition of the specific position of regions within European territory or to the placement of appropriate spatial development strategies in a transnational and European development context. The question remains as to which mechanisms within the concept of territorial cohesion could actually contribute to greater coherence between EU policies which have a territorial impact.

From the material presented so far, it is not clear how the concept of territorial cohesion will contribute to the consideration of Europe's great territorial and cultural diversity as an EU potential, or how the concept will take into account the importance of proportioning sustainable economic growth on the basis of territorial cohesion.

An elimination of the above-mentioned problems would provide space for the initiation of a wide range of potentials based on an efficient implementation of the territorial cohesion concept. Territorial cohesion as a theoretical concept incorporates a concern for the co-ordination of a comprehensive integrated approach and the concern of actively pursuing balanced development throughout the territory, which is also a concern of the regional economic approach.[20] This can be transferred into the political concept.

Territorial cohesion is meant to be a key for better horizontal co-ordination between spatially relevant policies and actions and for better vertical co-operation between the various tiers of government. In the ESDP framework, the EU, member states, and candidate countries Bulgaria and Romania, as well as Norway and Switzerland, finance the ESPON programme to encourage "further exchanges of experience and good practice, in particular through networking actions", and to support policy development.

[19] Peyrony (2005)

[20] Faludi (2004)

The concept of territorial cohesion can potentially contribute to sustainable and balanced development. Territorial cohesion, meaning the balanced distribution of human activities across the Union, is complementary to economic and social cohesion. Hence, it translates the goal of sustainable and balanced development assigned to the Union (Article 2 of the Treaty) into territorial terms.

The potential of the territorial cohesion concept can be seen as supporting the fulfilment of the principle of equality, as it includes fair access for citizens and economic operators to Services of General Economic Interest (SGEI), irrespective of the territory to which they belong (Article 16 of the Treaty). Important is its latent ability to support the activation of the unique development potentials of EU regions as well as to contribute to the definition of the position of the regions in the European territory and to the placement of spatial development strategies in a transnational and European development context. The concept of territorial cohesion can help to ensure a taking into account of the importance of promoting balanced and sustainable economic growth on the basis of the synergy between specific territorial potentials, as expressed in the quality of territorial cohesion.

It is expected, that the concept of territorial cohesion will contribute to a greater coherence between EU policies which have a territorial impact, and to a consideration of Europe's great territorial and cultural diversity as an EU potential. The concept of territorial cohesion can contribute to an increased competitiveness of the EU, as the territorial imbalances constitute factors of competitiveness. The concept can lead to the better consideration of sectoral and spatial components in the future EU Cohesion Policy, integrating the approach and coherence between national policy priorities and those for transnational and interregional cooperation.

Expectations linked to the cohesion policy differ among the various groups involved in this discussion. The expectations of specialists – practitioners and scientists in the field of spatial planning – are critical in their reactions to official political statements. On one hand, (optimistic) the potential effects of the implementation of a sophisticated territorial cohesion policy exist. On the other hand, realistic (or pessimistic) estimations of the effects based on current information about the territorial cohesion policy features from the decision making sphere are under discussion.

Politicians' officially expressed expectations can be briefly formulated as follows:

- Strengthening competitiveness and reducing disparities
- Improving the factors of competitiveness to promote territorial cohesion
- Facilitating structural changes throughout Europe
- Enabling regions to respond more effectively to the opportunities generated by the world's largest single market

- Differentiating between territorial imbalances representing factors of competitiveness and territorial imbalances representing handicaps (the distribution of population and wealth and the geographical handicaps affecting certain areas are conducive to territorial imbalances). The cohesion report analyses territorial imbalances, which, at the same time, could constitute factors of competitiveness, and which could correct the imbalances considered before).
- Safeguarding the equality of EU inhabitants concerning equal access and the efficiency of services (Art. 16 "[..] acknowledges that services of general interest (SGI) should, in particular, promote social and territorial cohesion, in order to provide equal access to SGI for all citizens wherever they happen to live or work in the Union").
- Developing a new approach for the policy of disparity reduction based on a new understanding of different kinds of disparities. As presented in official statements, the introduction of a territorial dimension of cohesion highlights the need to take the diversity of the European continent into account in order to achieve the rich and complex meaning of cohesion. Each territory has assets and also faces constraints that development policies must assess and integrate in order to be efficient.
- Integrating different spatial/territorial dimensions of cohesion. There is a need for a broader view of cohesion which encompasses many dimensions of territorial development, and the associated interactions reflected in the official documents of the EU. Politicians stress that the major challenge is to improve the coordination of sectoral and developmental policies which have territorial impacts.
- Improving the integration of European territories by reinforcing co-operation and networking among them.

In order to meet to expectations, the territorial cohesion policy has to be based on a clear vision of balanced spatial development, which does not yet exist in the enlarged EU. ESDP (EUREK) has not yet reflected EU enlargement and the main problems of territorial imbalance regarding regions in the new EU member states. The absence of such a vision can lead to inefficiency in political decision-making where there are no functioning concepts of territorial cohesion. Success is to be measured in the way in which the ensemble of territorially relevant policies has an effect on particular areas and the lives of people who live and work within them. It will not be possible if the territorial impacts of EU policies in areas such as transport, energy or environment are not immediately addressed. These policies should form an integrated part of a coherent whole, and spatial development policies at the national, regional and local levels should follow a set of principles that would underpin more sustainable and balanced territorial development within the Union.

References:

Akademie für Raumforschung und Landesplanung (ARL) (2004) Notwendigkeit einer Europäischen Raumentwicklungspolitik. Positionspapier aus der ARL no.60

Barroso J M, Verheugen G (2005) Working together for growth and jobs, A new start for the Lisbon Strategy, Communication to the Spring European Council, SEC (2005) 192, SEC(2005) 193, EC Brussels

Battis U et.al. (2005) Ausgestaltung der Raumentwicklung in der Europäischen Verfassung, Expertise zur Position der EU-Mitgliederstaaten zur künftigen Aufgabenverteilung zwischen der EU und den Mitgliedstaaten im Bereich der Raumentwicklung, Endbericht, Forschungsauftrag des Bundesamtes für Bauwesen und Raumordnung, Norr Stiefenhofer Lutz, Humboldt-Universität zu Berlin

BBR, Bundesministerium für Verkehr Bau und Wohnungswesen (eds.) (2005) Informelles EU-Ministertreffen zum Thema territorialer Zusammenhalt vom 20.-21. Mai in Luxemburg: Schlussfolgerungen des Vorsitzes

COM (2005) Lisbon Action Plan Incorporating EU Lisbon Programme and Recommendations for Actions to Member States for Inclusion in their National Lisbon Programmes, Companion documents to the Communication to the Spring European Council 2005 / Working together for growth and jobs, 24 final, EC Brussels

David C H (2005) Zur Konvergenz der nationalen Raumordnungspolitiken Frankreichs und Deutschlands im Post-EUREK-Prozess. In: *Raumforschung und Raumordnung*, Issue 1/2005 pp 11-20

ESPON Coordination Unit (ed.) (2005) In Search of Territorial Potentials, ESPON, Midterm results by spring 2005, [available on Internet: http://www.espon.eu/mmp/online/ website/content/publications/98/856/file_319/ESPON_synthesis_report_II.pdf]

ESPON (12.07.2005). ESPON 3.2; Spatial Scenarios and Orientations in Relation to the ESDP and EU Cohesion Policy (2004-06), Second Interim Report Part 1, [available on Internet: http://www.espon.lu/online/documentation/projects/cross_thematic/2705/sir-3.2_part1.pdf]

ESPON (12.07.2005) ESPON 3.1; Integrated Tools for European Spatial Development, Final Report Part A, Scientific and Policy Oriented Conclusions of ESPON Results Until September 2004, [available on Internet: http://www.espon.lu/online/documentation/projects/cross_thematic/2229/fr_3.1_part_a.pdf]

ESPON (12.07.2005) ESPON Project 2.3.2; Governance of Territorial and Urban Policies from EU to Local Level. [available on Internet: http://www.espon.lu/online/documentation/projects/policy_impact/2785/sir-2.3.2.pdf]

ESPON 1.1.1 (2004) The Role, Specific Situation and Potentials of Urban Areas as Nodes of Polycentric Development. Final Report. Stockholm: Nordregio – Nordic Centre for Spatial Development

ESPON 2.2.2. (2004a) Pre-Accession Impact Analysis. Third Interim Report. Institute for Regional Development and Structural Planning, Erkner

ESPON in progress (2004b) Preliminary results by autumn 2003. European Spatial Planning Observation Network

European Community (eds.) (2006) The Territorial State and Perspectives of the European Union Document: Towards a Stronger European Cohesion in the Light of the Lisbon and Gothenburg Ambitions 2006, first draft, Editorial Group 26, European Community, Brussels, June

EU Informal Ministerial Meeting on Territorial Cohesion (2005) am 20/21.05.2005 in Luxembourg, Presidency conclusions, Brussels

Europäische Kommission (eds.) (1999) Europäisches Raumentwicklungskonzept (EUREK): Auf dem Wege zu einer räumlich ausgewogenen und nachhaltigen Entwicklung der Europäischen Union. Angenommen beim Informellen Rat der für Raumordnung zuständigen Minister in Potsdam, Luxemburg

European Spatial Planning Observation Network (ESPON) (eds.) (2005). In search of territorial potentials. Mid Term Results by Spring 2005.

Faludi A (2004) Spatial Planning Traditions in Europe: Their Role in the ESDP Process *International Planning Studies*, Vol. 9, 2-3, 155-172, May-August 2004

Faludi A (2004a) Territorial Cohesion: A Polycentric Process for a Polycentric Europe. Working Paper Series 2004/03, Research team Governance and Places, University Nijmegen

Faludi A (2004b) Territorial Cohesion: Old (French) wine in new Bottles? In: *Urban Studies*, 41(7), pp 1349-1365

Faludi A (2004c) The Open Method of Co-ordination and 'Post-regulatory' Territorial Cohesion Policy. In: *European Planning Studies*, vol.12 / no.7, October 2004, pp 1019-1033

Faludi A (2005) Territorial Cohesion: an unidentified poltiical objective – Introduction to the special issue. In: Faludi (ed.) *Town Planning Review*, Special issue on European territorial cohesion, Issue 76 / no.1, pp 1-13

Faludi A (2005a) Polycentric Territorial Cohesion Policy. In: *Town Planning Review*, Issue 76 (1) 2005, pp 107-118

Finka M (2005) Competition and Cooperation – Classification of Instruments. In: Giffinger R (ed.) (2005) Competition between Cities in Central Europe: Opportunities and Risks of Cooperation, ARL/TU Wien/SPECTRA, Hannover, Vienna, Bratislava

Finka M (2003) The Future of Industrial Cities and Regions in Central and Eastern Europe. In: IoeR (eds.) (2003) The Future of Industrial Regions in Decline – Conversations between the U.S. and Europe, Dresden, pp 5-11

Finka M (2002) The Role of Planning in Increasing Ethical Behaviour. In: Petrikova D (ed.) (2002) Planning, Ethics and Religion II. FA STU Spectra Centre - ROAD, Bratislava, pp 146-151

Finka M, Petríková D (eds.) (2000) Spatial Development and Planning in the Context of EU Enlargement, Central European Training Centre in Spatial Planning, Bratislava/ Hannover/ Newcastle/Grenoble

Finka M, Petríková D (2002) Achieving Social Equity Through Effective Spatial Planning Policies and Tools. In: Petríková D (ed.) (2002) Planning, Ethics and Religion II, FA STU Spectra Centre - ROAD, Bratislava, pp 192 – 205

Gestel van T, Faludi A (2005) Towards a European Territorial Cohesion Assessment Network. A bright future for ESPON? In: *Town Planning Review*, Issue 76 (1), pp 81-92

Gustedt E (2005) Territorial Cohesion of Europe – From a Vague Notion, to Clear Idea, AESOP Conference, Vienna

Hall M, Ache P et al. (2006) ESPON 2.2.1, The Territorial Effects of Structural Funds. Third Interim Report, 2004, Nordregio, Mcrit, EPRC, infyde, itps, Systema, [available on Internet: http://www.nordregio.se/espon2.2.1.htm , 1. 5.2006]

Commission of the European Communities (2004) Employment, economic reforms and social cohesion – towards a Europe based on innovation and knowledge, Document from the presidency 5256/00, Lisbon/Brussels 2000

Laissy A P (2004) Ministerial Meeting on Territorial Cohesion on 29 November 2004 in Rotterdam CEC 2004

Lévy J (2003) Aménagement du territoire. In: Dictionnaire de la géographie et de l'espace des sociétés, Paris - Berlin

Lévy J, Lussault M (eds.) (2003) Justice spatiale. In: Dictionnaire de la géographie et de l'espace des sociétés, Paris - Berlin

Malchus von V (2005) Strategische Leitbilder für die EU-Kohäsionspolitik, Manuscript

MKRO, Hauptausschuss der Ministerkonferenz für Raumordnung (eds.) (2005) Fortentwicklung der Europäischen Raumentwicklungspolitik, 28 April

Office for Official Publications of the European Communities (eds.) (2004) European Communities/ DG Regional Policy, Interim Territorial Cohesion Report (Preliminary results of ESPON and EU Commission studies). Luxembourg Office for Official Publications of the European Communities

Office for Official Publications of the European Communities (eds.) (2004a) Interim Territorial Cohesion Report, Preliminary results of ESPON and EU Commission studies, DG Regional Policy, Luxembourg

Perony J (2005) Territorial cohesion: evolutionary background, implementation in the frame of 2007/2013 EU policies, AG TC ARL - working material, ARL Hannover

Rawls J (1971) A Theory of Justice. Oxford University Press, Oxford

Schön P (2005) Territorial cohesion in Europe. In: *Planning Theory and Practice*, Issue 6/3, pp 387-398

4 Regional Disparities – Reasons Enough to Run a Cohesion-Oriented Spatial Policy?

Hans-Peter Gatzweiler, Wendelin Strubelt

4.1 Spatial Equity – old and new Discussions

In the context of the Federal Republic of Germany, the creation of equal living conditions has been one of the major goals and targets of spatial development and policy for decades. In the Spatial Policy LawFederal Regional Planning Act of 1965 it is set as the major goal. Nowadays, it is included as an overarching aim in a revised version, namely, as sustainable development. Recently, however, an intensive public discussion has arisen about the question of whether it still makes sense to follow the target of equal living conditions. Oddly enough, this started as a result of an interview by the president of the Federal Republic of Germany, Horst Köhler[1], who is a trained economist. He raised the question of whether it was good for economic and, in the end, for political reasons, to maintain a similar infrastructure in every region, which means in very sparsely populated areas as well as in agglomerations, or whether it was good even to try to reach this level.

The public uproar caused by this interview shed light on the strong position of the idea of equity in living conditions as a political aim, even if it has only been some form of political phrase used in public and not just within the context of experts, sometimes without considering the concept's implications. In general, this was a revival of a discussion which has been going on a long time and was also lively in the 1970s[2]. At that time, however, the realistic view of this goal did not prevail. The supporters of this goal, backed by a league of spatial development experts, prevailed and it remained an overarching goal which as a whole is unrealistic. As some argue, however, the way of approaching the goal is the intention and not the goal itself in its pure form.

---

[1] The president of the Federal Government of Germany, Horst Köhler. Interview with the journal *Focus*, Issue 38, 13 September 2004

[2] Kommission für wirtschaftlichen und sozialen Wandel (1977) chap. 6, "Raumordnungspolitik Neuordnung der Funktion der Räume", especially the sentence on pg. 323: *The overall aim of spatial policy is the creation of equal living conditions in every region of the Federal Republic of Germany, which was the aim through the concept of central places which are evenly distributed throughout the country, that means 650 so-called medium central places within the context of the spatial programs of the "Länder," which is supposed to be reached but can not be realized due to financial, ecological, technological and social economic reasons.*

Within the discussion about equalized functional regions, on the one hand, and regions characterised by specialties or by some sort of division of labour, which even implies unequal standards of living measured by standard indicators, on the other hand, the question was raised about whether the ubiquitous equity can be maintained at all in future. The argument that we should orient spatial development policies more towards spatial specializations or divisions of labour between regions was also developed, which could finally lead to easily recognizable and perhaps even accepted differences when compared to the circumstances or living conditions in other regions. For the individual person this would even create the opportunity to make choices by moving, i.e., by reacting to perceived inequalities regarding spatial mobility. It would also provide the opportunity for every individual to have some sort of standard regarding infrastructural services in the region in which he/she is born or is living in. There would not, however, be a ubiquitous, high level of infrastructural services in all parts of the country because there are regional differences and regional disparities in relation to living conditions, which offer advantages or disadvantages according to regional differentiation, and about which everybody can deliberate upon depending on the impact such differences have on them. Everybody has to make his or her own choice. It will not be made by the state for them.

This discussion about the concept of the equity of living conditions and a possible ubiquitous distribution or implementation in all regions was attenuated in the 1970s, together with the ideas and forces backed by the experts of spatial policy. Spatial equity remained the general goal although it was quite clear that it would never be reached. It therefore has to be considered a political credo rather than a realistic goal. Since German reunification, however, and on account of the completely new and different spatial situation – a very densely populated part of Germany was united with another part that had a much lower population density, and which turned out to suffer from depopulation – deindustrialization and fiscal problems involved when maintaining this equality have since been quite evident. This former policy approach and the question of how it could be maintained (if it could be maintained at all) appeared under a completely new light. We have to face the fact that in former times it was possible to maintain equality while helping regions faced with less favorable conditions, on the one hand, while on the other hand helping growth regions to finance their adjustment policy forced by such growth. Due to a lack of money there is now the problem of choice in regard to the formation of these policies. And the choice is a question of the allocation of rare resources by politicians and has to reflect the possible and different implications of these policies of equalization, i.e., which regions will be supported and which regions will be less favoured. Thus, we once again have intensive discussions about the question of whether the use of an equalization policy to deal with regional disparities can still be financed.

This new discussion about the quality of living conditions has to consider the current situation of regional disparities and the perspectives of possible future or expected developments. In addition, all attempts to accept the goal of equal living conditions as some sort of a general or even dogmatic doctrine, in itself a taboo within the context of spatial policy, could have severe and problematic implications for the situation of spatial policy itself. When observing the differences between the materialization of the goal in relation to reality and especially in relation to the continuous persistence of disparities over the course of time, the question could be raised as to why the degree of implementation of spatial policy is comparatively lower than in other policy areas – a constellation which might question the future value of this very policy by demanding much more than it is able to implement. If we think it through, however, we should not neglect to mention that the existence of this goal of equal living conditions in every region has had an impact on regional development policies. Equal living conditions are therefore a difficult goal, and seldom clear cut, which can be proven by simple analyses. They have to be measured by very complex spatial constellations, which on account of the difficulties of comparing almost incomparable spatial constellations, is a very complicated conundrum.

In addition, we have to face the fact that most spatial developments, regardless of their negative or positive direction, can only be changed over a long period of time. They are based on long-term trends, which develop continuous forces beyond the short-term perspectives of political activism and public awareness. In order to understand these trends and the facts behind them, we need an analytical approach with a high level of observation and analysis; this means some sort of continuous and consistent analysis within the context of a spatial information system which continuously observes and analyzes spatial and urban developments. The Federal Office for Building and Regional Planning is required to maintain this by way of the Federal Regional Planning Act.[3]

4.2 Why have a Spatial Policy oriented toward Equity?

A spatial policy oriented toward equity faces two big challenges. **Firstly** we have to face the fact that European integration and the globalization of the economy have led to an intensification of competition in all areas and markets, resulting in growing competition between different locations on a regional and international basis. As a result, we can observe rapid economic change which will lead to a constellation of economic growth in some areas together with economic decline in other areas, producing a high level of unemployment. All this creates heavy pressure on the respective regions, thus producing regional disparities, which is the opposite of the overall aim of spatial and social equity. **Secondly** we have to face the fact that an enormous demographic change in German regions is occurring on account of a declining population and a growing segment of elderly people. These constellations

[3] The following analysis is based on different articles within the context of the issue "Gleichwertige Lebensverhältnisse" in the journal *Informationen zur Raumentwicklung*, Issue 6/7.2006

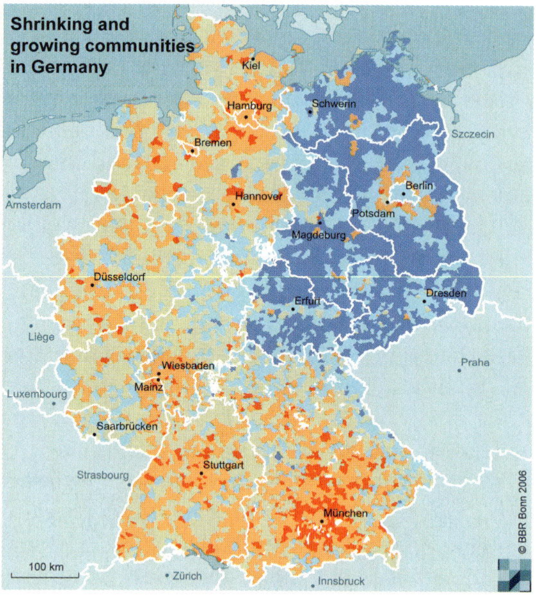

Circle of shrinking and growing

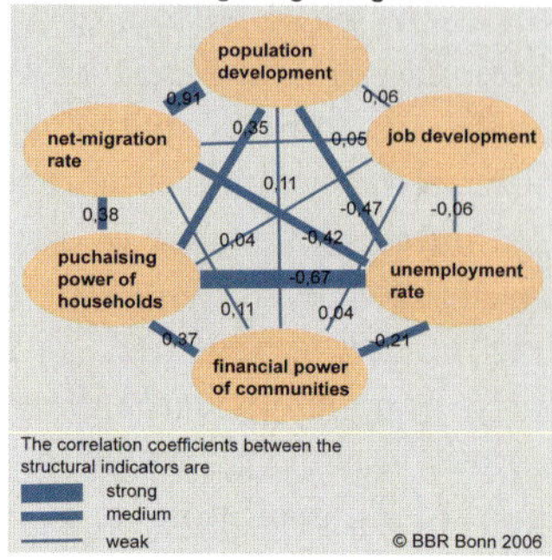

Considered structural indicators:

- population development 1997-2004
- net-migration rate 2002/2004
- job development 1997-2004
- unemployment rate 2003/2004
- communal financial power 2003/2004
- purchaising power of households 2004

- strongly shrinking
- shrinking
- stable
- growing
- strongly growing

Class definition according the frequency of indicator values within the bottom quintile (< 20% of all values) and within the top quintile (> 80%of all values):
- racily shrinking: 4 - 6 indicators within the bottom quintile
- shrinking: 1 - 3 indicators within the bottom quintile
- stable: no indicators within the bottom od top quintile
- growing: 1 - 3 indicators within the top quintile
- racily growing: 4 - 6 indicators within the top quintile

Source: Spatial Monitoring System of the BBR - communities

Contentment of people with their place of domicile in shrinking and growing communities

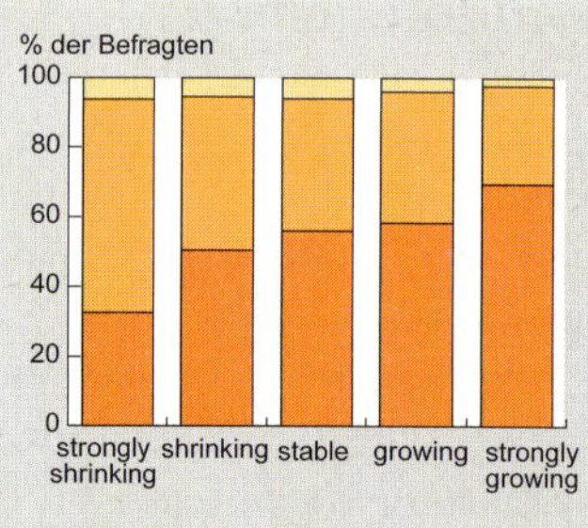

"Question across-the-board: How content are You with Your city or municipality? Please indicate Your contenteness to a value scale from 1 to 7. Value 1 means that You are "very discontent", value 7 means, that You are "very content". The values in between are to express different shades of Your contenteness."

content: scale values 6 and 7
partly: scale values 3 to 5
discontent: scale values 1 and 2

Sum of the respondents of the BBR survey of the years 2002 to 2004

Class definition see map above

Source: Annual Houshold Survey of the BBR, Spatial Monitoring System of the BBR

Fig. 1: Regional differences in living conditions. Source: Spatial Monitoring System of the BBR ("Bundesamt für Bauwesen und Raumordnung")

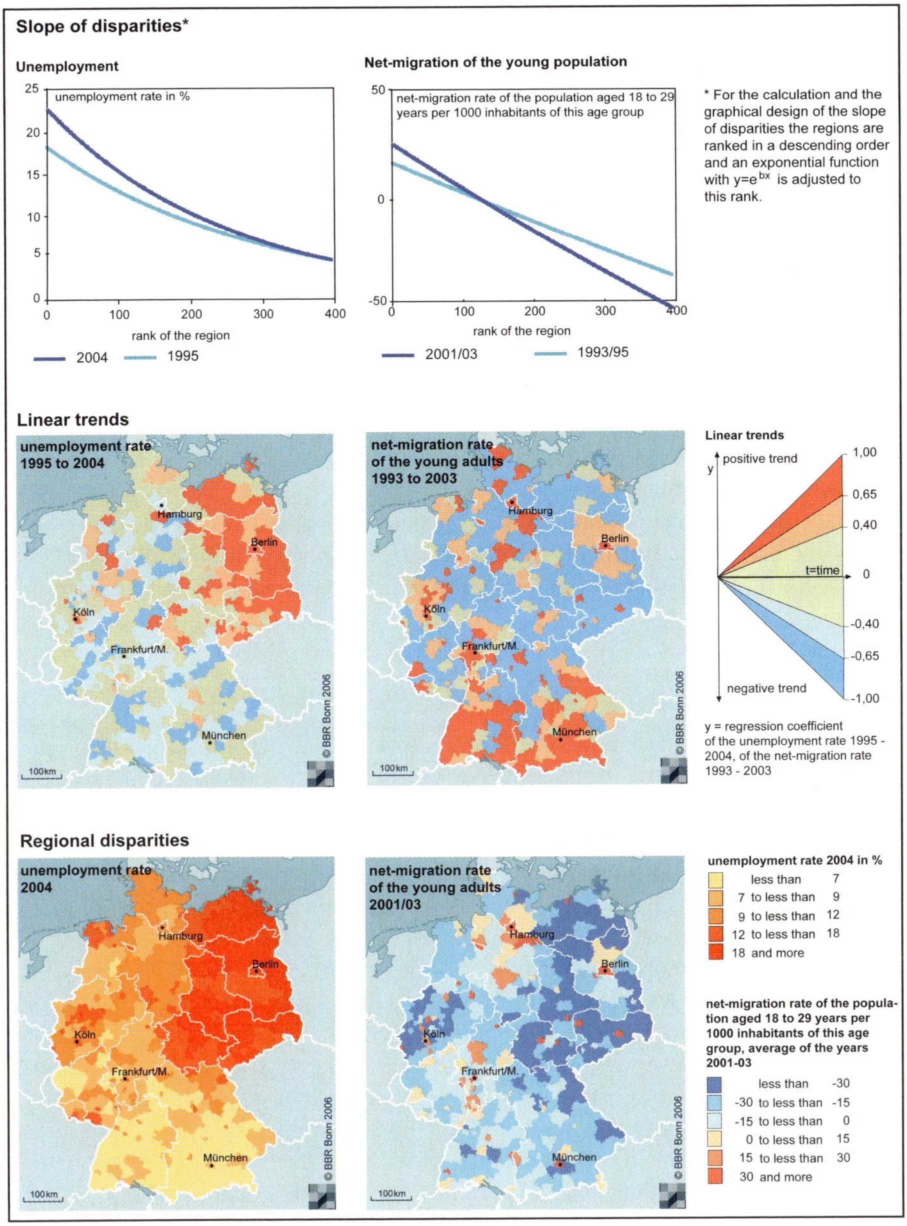

Fig. 2: Regional disparities with regard to employment and their development. Source: Spatial Monitoring System of the BBR ("Bundesamt für Bauenwesen und Raumordnung"), modified NUTS 3 regions of date 31.12.2003.

challenge a modern welfare state regarding its duty to guarantee a good standard of living, and create heavy pressure on institutions while demanding a new level of adaptation.

Both challenges have led to new regional disparities throughout the country and to various constellations in Germany regarding spatial and urban development, featuring processes of growth and of shrinkage in different but sometimes almost in nearby places at the same time. This can be easily seen if we classify different communities according to indicators which combine such aspects as the development of population and migration, the development of jobs, the level of unemployment, regional differences in communal financial power and the purchasing power of households. The choice of these combined indicators reflects the fact that processes of shrinkage and of growth are subject to multidimensional and systematic processes. We consider growth or shrinkage to be systematic, consistent phenomena, whether or not regressive negative processes or respective positive developments are dominant and reinforce each other.

A look at the results of our analysis presented in Figure 1 reveals that shrinkage and growth are quite unequally distributed. In the east of the republic there is a heavy concentration of shrinking cities and communities while in the west we find ones exhibiting growth. And by looking at the diagram featuring the contentment of people with their place of domicile in shrinking and growing communities, it is easy to see that there is a strong connection between being content and the situation in which people live. In heavily shrinking communities only a third of people questioned are content with their living conditions in the place they reside in, quite in contrast to about 70 percent in places where growth is occuring. We can thus state that the situation of a place belonging to an area exhibiting either growth or shrinkage is very well perceived by residents and has a great impact on their level of contentment. For future spatial development this can be decisive, because it might influence the choice of location and decisions regarding the future spatial behaviour being a decisive momentum for migration.

From different analysts we know that in general two factors, respectively indicators, are the best explanations for regional differences regarding the judgment of living conditions: the rate of unemployment and the net migration rate of young adults (18 to 30 years of age) in relation to their percentage of population, as well as the population in toto. The unemployment rate has an especially great impact because we know that in regions where this rate is high, contentment with living conditions is very low. In these cases the possibility that young people migrate is high because migration depends on the chance of getting good vocational training and the possibility of finding employment. Both facts are highly interconnected.

All these facts show that policy action is needed in order to achieve a spatial policy of equity, because the creation of sufficient jobs is without question the central dimension of equal living conditions. A lack of jobs is typical for declining regions, and is the result of structural deficits in economic development. While the process of transforming a formal, centrally guided national economy has resulted in a structural landslide in the east, we have similar problems in the west due to structural crises stemming from the aging of some industrial regions which are not able to compete in growing global markets.

4.3 Regional Disparities with Regard to Employment

Our analysis of those regional disparities concerned with unemployment not only presents the latest information with regard to the year of observation but also presents a comparison of the development which started in the mid-1990s (see Figure 2). This clearly shows the sharp regional differences in this development and it shows the crucial position of regional disparities with regard to employment and development for the present and future German situation. The slope of disparities shows the dramatic increase in regional disparities with regard to unemployment. Looking at the slope of disparities and the ranking of regions in the course of time, i.e., between 1995 and 2004, it is also possible to observe quite different situations because some regions were able to reduce their rate of unemployment. In other regions, however, the level of employment increased. All in all, we can observe that disparities between different regions with regard to unemployment have been growing. This means the slope of disparities is deeper now than it was in the mid-1990s. If we turn to another important employment indicator, the net migration of the young population (aged between 18 and 30), we can also observe a growing level of regional disparities on the national level.

Maps indicating the development of the unemployment rate between 1995 and 2004, and of the net migration of young adults between 1993 and 2003, present a clear picture about the regional development of these trends (see Figure 2). At the same time we can see that between 1995 and 2004, there was a simultaneous growth and decline of unemployment in different regions, which, seen from a national comparative perspective, resulted in a growth of regional disparities. And if we compare the net migration of young adults to unemployment, this constellation is proved. On account of an even worse situation in vocational training and the possibility (or impossibility) of getting a job, more and more young people leave their places of birth. This is especially the case in the so-called "neue Länder" (new federal states, the former GDR) but also in some parts of western Germany. And when these young people do migrate, the most popular destinations are the big cities in the west, and to some degree in the east as well.

All in all, this clearly signifies that in toto, Germany has not developed a more homogeneous regional structure. Instead, we can observe sharp differences between old industrialized regions and booming service-oriented regions on the one hand. On the other hand, we still have the sharp differences, if not cleavages, between east and west which started with German reunification. We can nevertheless

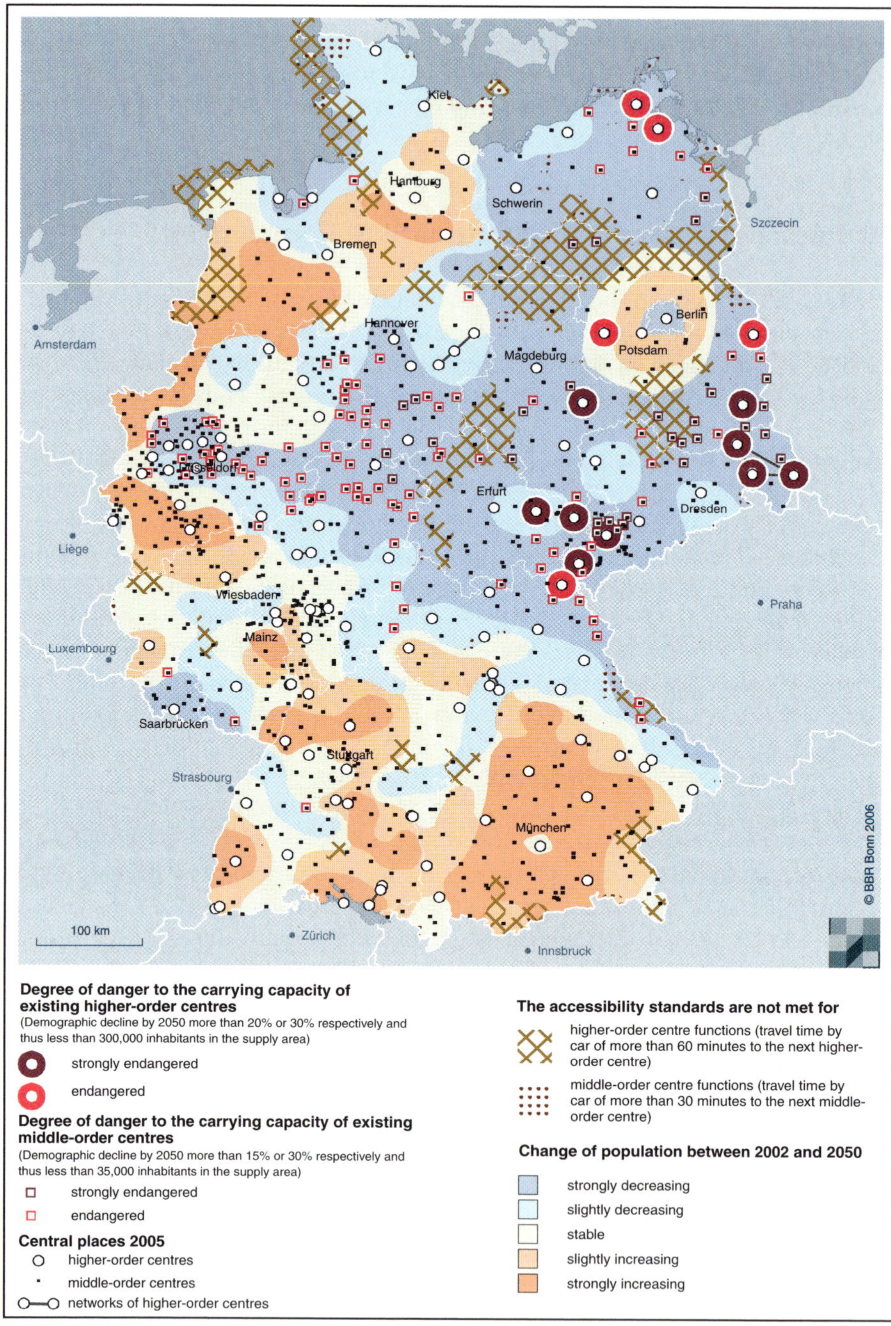

Fig. 3: Carrying capacity and accessibility of central placest. Source: Accessibility Model of the BBR ("Bundesamt für Bauenwesen und Raumordnung"), population prognosis of the BBR 2002-2050/Exp

observe some processes of differentiation in the east, in the new "Länder", because the attractiveness of cities such as Leipzig, Dresden or Jena, not to speak of Berlin, creates a new disparity between rural and urban regions in this part of the country as well.

4.4 Regional Disparities with Regard to Services for the Public

Another central dimension of equal living conditions is the supply of services with regard to daily life. Within the German context, one of the mostly used political instruments to develop a spatial policy in regard to evenly distributed and sufficient services for daily life is the "Zentrale-Orte-Konzept" (*central places concept*). In order to evaluate the functioning of central places in granting the establishment of institutions for everyday life, we have selected two main criteria: firstly, the amount of a sufficient number of people living in order to get in contact, that means enough carrying capacity, and secondly, the accessibility of central places.

The analysis of the carrying capacity and of the accessibility of central places is based on the so-called system of central places which is supposed to guarantee a sufficient supply of everyday services. This system differentiates between lower-order, middle-order and higher-order centres and was created by political agencies concerned with spatial development. The system of central places is once again based on special programs and plans in the federal states ("Länder"). Medium-sized centres were analyzed with regard to their carrying capacity within their functional region. In our analysis, carrying capacity was said to be problematic or greatly endangered if a central place exhibited a population decrease of between 15% to 30% by the year 2050, which would mean that it would drop below the threshold of 35,000 people within their functional region. Higher-order centres were also analyzed with regard to their accessibility by car according to their carrying capacities within their functional region. The carrying capacity was said to be problematic or even strongly endangered if such higher-order centres have a population decrease of more than 20% or 30% respectively and would therefore drop below the threshold of 300,000 people within their functional region. Regarding problems or deficits in accessibility, we said that deficits exist in those regions where it takes more than 30 minutes by car to reach the next medium-sized centre or more than 60 minutes to reach the next higher-order centre.

Our analysis shows the consequences of a future population decline quite clearly. On account of demographic development quite a lot of medium-sized and higher-order centres are in danger of losing their carrying capacities (see Figure 3). This is also the case in regions where a dense network of central places is prevalent, because in these cases the carrying capacity might be in danger as well in the future. What are the consequences? According to our analyses, we think that many public services and the standards of services for everyday life have to be analyzed with regard to their need, on the one hand. On the other hand, we have to evaluate the different steps of order between the central place system and possibly modify them. This means that in some places we will have to see if some services can still be supplied or whether they have to be transferred to other higher-order centres. All in all, this means evaluating the problem of whether the almost ubiquity of some services can be maintained, and if society can still afford them.

With regard to standards of accessibility and according to our analyses, it seems that the current central places system fulfills the basic requirement of accessibility within 30 to 60 minutes when driving by car to a large degree, on the Federal level. Some deficits exist in regard to coastal areas and some peripheral locations, in the context of the old Federal Republic, but all in all they can be compensated for by the current central places system.

If we look at the different level of deficits in the new "Länder" (the former GDR), however, we can see that the existing central place network does not fulfill the requirements of accessibility and carrying capacity anymore (see Figure 3). The standards required by this system are not met. The question will be raised as to whether these standards can be reached in the future or if the standards will have to be lowered. In some regards this would require a rethinking of the level of equity that society is able to supply in the future. This could end up in a process of rethinking, by defining minimal standards which have to be supplied everywhere, and higher standards, which have to be supplied in a more structured way, i.e., not everywhere.

4.5 Conclusions concerning Spatial Policy

In general, we can say that in the future a spatial policy oriented toward equity will face two big challenges: namely, the need to create enough employment opportunites in all regions and the ability to guarantee a proper and affordable standard of living accompanied by the necessary everyday services. As mentioned above, both challenges are the results of two basic changes in the economy and society: globalization and demographic change. Globalization means an intensification of competition in all markets, with the consequence of growing international competition. At the regional level it has serious impacts and results in rapid economic change and will probably create a continuous high level of regional unemployment. Demographic change in Germany has impacts on the shrinkage and aging of the population, especially in low density regions which have strongly declining populations. Public infrastructure and the supply of everyday services will be endangered.

Both challenges or developments influence each other, and they interact with each other. Spatial development is influenced by the population, the presence of jobs and infrastructure. People will migrate in order to get jobs in prospering regions and firms are looking for locations in regions which are economically promising for them as well. Infrastructure is concentrated in regions where the demand from the economy and the population is the largest. Other regions which are suffering a loss of population, and firms, will lose their attractiveness with regard to infrastructure, which again influences the migration process as a consequence. These interactions are followed by an accumulation of regionally problematic situations. This may start a vicious circle: a lack of employment, economic structural problems, and a scarcity of services for everyday life is concentrated in some regions and the phenomena strengthen each other. As mentioned before, the reason for this has to be ascribed to global economic change and to national and regional change.

The target of spatial policy is to create equity in times of a concentration on economic competitiveness and growth targets, however, it has to be a policy which, within the context of a sustainable development strategy, tries to diminish the largest differences in development, especially with regard to unemployment and the structure of services for everyday life. This means that in every region there should be a minimal level of standards in both areas, which in the end means taking the risky way between development and equalizing strategies at the same time. We need strategies for steering the rapidly developing, booming (urban) regions as well as strategies for helping regions compensate for their deficits.

On June 30th 2006, the German Standing Conference of Ministers Responsible for Spatial Planning agreed on new "Leitbilder" (concepts) and action strategies for the future spatial development of Germany. There are two concepts: growth and innovation, and the guaranteeing of everyday services in all regions. This means a new orientation with regard to the understanding of spatial development. The aim is to strengthen the orientation toward growth and innovation by spatial policy in order to create chances for new employment. There is also an orientation toward a new design for the equalizing strategy, by changing or strengthening carrying capacity and the accessibility of central places within the context of the central places system.

This policy can be characterized as an approach to maintaining a spatial structure or a national basis whose competitiveness is measured by global standards or developments, and not just measured by the needs and demands of the people. It is a policy which in some way results in conflicting ends. And finally, it is also a question of affordability. West Germany was able to tackle these conflicting targets for a long period of time. After German reunification, the burden on society has become heavier and the need to rethink some standard approaches is now in question. Choices have to be made which, for the moment, might result in inequity. The hope is however, that in the end the support of booming metropolitan areas will help to promote declining regions as well. It is a policy of structuring aid, not a policy of sheer distribution.

5 Territorial Cohesion, Brain Drain and Digital Divide

Hans-Joachim Bürkner, Ulf Matthiesen

Introduction

Declining regional economies often trigger out-migration of the local population. Characteristically, this process is highly selective with regard to age and education: It is the best educated and youngest people who migrate first, leaving relatively disadvantaged, older-aged and immobile segments of the population behind. This gradual decline in the qualified workforce has become known as *brain drain* during the past few decades. In eastern Germany the general pattern of brain drain processes has been reinforced since 1990 due to a clear gender bias brought about by an increasing number of young women on the move. Additionally, severe and continuous population losses culminate in a characteristic cascade of economic crisis, e.g., rising unemployment, decreasing investment rates in traditional branches, and decreasing private and public sector investment into primary and secondary education and vocational training.

Nevertheless, there is no compelling evidence to explain brain drain exclusively in terms of economic decline. Interdependencies between the economy and migration tend to develop a logic of their own, rendering a causal explanation almost indeterminate concerning the direction of possible cause-effect schemes or push and pull factors. Although initially affected by economic crisis, inadequate or inapt human resources themselves (human capital) contribute much to a preservation or acceleration of peripheralisation processes. Regions structured by brain drain experience a dramatic decline in attractiveness, not only for people but progressively for domestic and global investors. This especially holds true for eastern German peripheries such as the German-Polish border region. Reduced investment by private agents and the relocation of enterprises both contribute to a blacking out of whole regions from the mental and agency maps of potential future investors – especially of global players. While many peripheries have thus fallen into the slipstream of global investment campaigns, European metropolises have experienced the positive effects of a concentration of capital and workforce, especially in newly expanding branches of the knowledge economy. These disparate trends pose questions about the present condition of territorial cohesion as well as about its future development.

5.1 Brain Drain in Eastern Germany – Increasing Governance Paradoxes related to the Outcomes of Post-socialist Peripheralization

Brain drain is a continuous, self-reinforcing trend in eastern Germany which became established early in the 1990s as a consequence of the abolition of the former socialist industrial base. In the following period, this trend led to the strong and continual out-migration of residents, which was mainly due to a threat of unemployment. Three particular groups have shown increased tendencies to move: young people, qualified and well-educated residents, and women. Moreover, there are often intersections of age, gender and qualifications, with a strong bias on gender: especially young women often leave the rural areas, and they leave peripheral regions in particular. In this way, the process of migration commonly observed in industrial societies (i.e. the high mobility of young males) became the opposite in eastern Germany. Between 1991 and 2002, 243,700 males and 447,600 females left eastern Germany.[1] Dominant motives for out-migration are the quality and standard of workplaces and options for professional qualification and learning. The goal of receiving a good professional education, of becoming qualified, of having a job perspective and encouraging work-life-balance are the main factors in the decision to migrate. In addition to *'rational factors'* there are also strong *'emotional factors'* – related to the raising of children, to social and local identities, and to the quality of a specific place, etc.[2]

The 'other side' of this collective decision process is the lack of decision made predominantly by males with lower competencies, lower mobility tendencies and less cultural heterogeneity. The relatively high concentration of these remaining persons with restricted capacities has two major effects: **1.** It enhances the co-presence of human resource deficits, especially in peripheral regions, and helps to sustain high long-term unemployment rates, which, in turn, effectuate problems of territorial and social cohesion, and **2.** It results in tightly knit social networks 'at home', which have an ambiguous nature. These networks, on the one hand, contribute to a sealing-off of local milieus from external stimuli, thus preventing the flourishing of creativity, open-mindedness and innovative thinking which are needed for a successful adaptation to new challenges. On the other hand, they ensure an extraordinary richness of local activities and place-making processes which can be filed under the heading of *'post-socialist localism'*.

In order to illustrate the spatial selectivity of theses processes, it should be mentioned that peripheral regions are affected in a disproportionately negative way. In a research project involving 250,000 national servicemen and women, Ebendrett, Hansen and Puzicha[3] recently showed that in particular the peripheral regions in eastern Germany are experiencing severe ongoing losses of human resources –

[1] Dienel (2005) and Haase (2005)
[2] Dienel (2005)
[3] Ebendrett/Hansen/Puzicha (2002)

resulting in corresponding deficits in levels of professional competencies, interaction dynamics and knowledge cultures.[4] One impending danger here is that in some of these regions the lowest level of competency required to engage in creative and innovative activity has nearly been reached. If local and regional competency level falls short of this minimum rate of capacity, it will not be possible to turn human resource losses into a brain gain for a very long time. Eastern German peripheries additionally run the risk of sliding into a "knowledge societal competency trap."[5]

The chronic crisis of Federal and "Länder" (states) fiscal systems has aggravated these dangers considerably – even though most of the five eastern German state governments are doing their best in trying to reduce expenditure in knowledge, learning and research-based areas as little as possible. For the moment being, however, a stabilization of state-side expenditures ("Überwälzung") seems to be the greatest possible success here.

This might not be good enough, however. With the exception of extra-university research (Leibniz Community, Helmholtz Community, etc.) the landscape of knowledge, research and higher learning in eastern Germany clearly falls short in comparison to that in western Germany. Instead, the political agenda generates truly utopian plans, e.g., the federal government should transform eastern Germany in its entirety into an Innovation Region – which would require considerable federal (!) investments in this field.[6]

Apart from the fact that the federal state's growing fiscal problems would quickly render such an initiative obsolete, a de-contextualized innovation initiative like this one, focusing only on learning and higher education, would aggravate the problems which the Innovation Resolution proclaims to solve. The reason for this unintended consequence is simple to see: the now better-educated people from eastern Germany would certainly take advantage of their new chances in more extended transregional labour markets and strive for better jobs – most of which are located in the west (western Germany, western Europe etc.). Out-migration and the loss of human resources would increase due to an improvement in human resource levels – leading to a severe governance paradox in the field of education and knowledge politics.[7] From this we may learn that traditional department-oriented policies regularly lead to new governance paradoxes: brain drain is increased either by the enhancement of educational mono-structures or by the arbitrary introduction of isolated educational structures into 'abstinent' environments. The latter case can be observed throughout eastern Germany,[8] especially in Brandenburg's state policy of letting scientific UFOs (small specialised universities, research institutions) touch down in thinly populated and economically devastated areas.

[4] Ebendrett et al. (2002)
[5] see Matthiesen (2004)
[6] SPD-Parteivorstand (2004)
[7] Matthiesen (2003)
[8] cf. Franz et al. (2002)

5.2 Territorial Cohesion: An outdated Notion?

When it comes to assessing spatial development trends in Europe, there is still much empirical work to be done in order to document and explain socio-economic inequalities, and to draw conclusions concerning changes in terms of territorial cohesion. And it is certainly not easy to validate assumptions about 'causal' relationships between territorial cohesion and socioeconomic inequality. One reason for this is that currently an enraged debate about social and economic inequalities and polarizations on the regional, national, European and even global level is under way.[9] A first result of this is that disparities within nation states have to be clearly separated from generalizations about disparities between nation states. Whereas poverty declined significantly in developing countries, etc. over the past twenty years, inequalities within nation states belonging to the enlarged EU increased considerably, leading to a new geography of income inequality.[10] Even territorial imbalances in the 'old' European Union show corresponding discrepancies between the evolution of polarization and regional inequality.[11]

Political reactions within individual nation states almost inevitably try to establish and defend the idea of territorial cohesion. At first glance, this idea appears to be the outcome of an older logic of spatial development, or, as it were, the spatial dynamics of the First Modernity. Up to the 1970s, the expansion of industrial production, urbanisation, suburbanisation, and the respective territorial arrangement of governance principles within the welfare state formed a tight structural intertwining. During this era, the paradigm of spatial policy referred to a balanced growth of centres and peripheries alike. It seemed to be a mere matter of time before spatially equal and just living conditions were achieved everywhere. A vision of the dissolution of spatial disparities was conceived as a linear outcome of growth-based modernisation. The diversification and flexibilization of production models after the end of the Fordist era, however, gave way to much more complicated patterns of growth and decline. The Second Modernity, in particular, has disclosed an increasing fragmentation and splintering of territorial relations. Economically declining and growing regions appear side by side on diminishing scales, forming increasingly subdivided mosaics of divergent development patterns.[12] Most of these relationships were also present during the First Modernity, albeit on a larger scale and with more chances for efficient redistributive policies. During the past decade, however, spatial fragmentation caused by the growing dynamics of globalisation, knowledge-based economies, and regionalisation has been fostered rather than mitigated by national and regional policies. This process is accompanied by a new political rhetoric of helplessness, if not fatalism, in face of the ubiquity of globalisation and the hollowing out of the nation state.

[9] see Ezcurra et al. (2006); Berger (2005); Milanovic (2002) and Milanovic (2005)
[10] Berger (2005) pp 269 ff
[11] Ezcurra et al. (2003)
[12] for Germany see Gatzweiler/Irmen (1997)

On the one hand, national politics propagates a return to universal growth and–as a sub-text–to patterns of coherent macro-regional development, e.g., as promised by the rise of the EU's 'blue banana' development belt. On the other hand, local and regional agents are imagined as being subdued to globalisation's all-dominant imperatives of action. Their divergent reactions are envisaged as more or less apt adaptations to new challenges, making micro-regions oscillate between helplessness ('governance') and strategic optimism ('empowerment'). As a result, the political view of eastern Germany's spatial development shows a schizoid divide. The perspective on hoped-for growth has different implications for relevant models of spatial development than the concurrent perspective on new disparities and their weak controllability: While the growth assumption triggers conventional ideas about achieving macro-regional homogeneity and large-scale compensation for disparities, the fragmentation assumption suggests new differentiated, flexible, hitherto untested strategies of small-scale development.

Consequently, the question of territorial cohesion becomes more virulent than ever, mainly in a normative respect – not as an abstract principle of spatial order but rather as a question of the socio-spatial connectivity of a society which has ventured into new differentiations and divisions, i.e., into the plurality of heterogenous lifestyles, the hardening of social distinctions, the exclusion from job opportunities for an ever-greater part of the population, and the abolition of old solidarities,[13] etc. Recent apologetic public debates about the future of eastern Germany's peripheral regions contain large amounts of 'tacit' scepticism: doubts about the capacities of inclusion available for a globalised society mingle with the search for alternatives to modernisation policies which have regarded post-socialist transformation as a historical defect rather than a positive opportunity for political fashioning and open-minded restructuring. This search, after all, has become more and more entangled within discussions about the evolution of the knowledge-based society and the options it might hold for the future development of peripheries.

5.3 Knowledge and the Periphery

During the 'long decade' from 1990 until 2005, it became popular among economic politicians and education planners to support knowledge-based development processes as being the clue for a development strategy, especially for peripheral regions. Relocating universities, research institutions, facilities for vocational training etc., to peripheral towns, and supporting clusters of small and medium-sized enterprises outside of agglomerations became a common political issue, in particular as part of eastern German regional policy.[14] Endeavours to create a counterbalance to peripheralisation, however, increasingly became foiled by the actual migrational

[13] The socio-political contracts of generations, for example.

[14] Sautter (2004)

behaviour of the population. Brain drain was and still is advancing. This means that poor regions had to bear a considerable part of the costs of educating and training the workforce, in favour of prosperous regions, without getting fair compensation. With respect to regional pathways of development, two variants are discernible here:

1. A vicious circle of peripheralization begins: outflows of educated workforce lower the significance of qualification as a locational factor, the region becomes less attractive to investors, local enterprises might increasingly feel a need to relocate because of a shortage of human capital, job losses and rising unemployment lower the motivation of young educated people to stay, and the numbers of out-migration remain high or even increase. If increasing endeavours are made to strengthen the educational basis, a one-sided alimentation of prosperous regions is fuelled.[15]
2. The co-ordinated education of highly specialised professionals around existing hi-tech clusters, universities, private research institutions, etc., at 'promising' locations might lead to a temporary stabilisation. Political strategies, however, often underestimate the enormous mobility of experts and professionals who are courted by headhunters at a global level. Thus, many of those educational clusters established recently (e.g., following the campus model of hi-tech, state-funded research and education) may turn out to be mere temporal clusters within quickly changing, fluid *'KnowledgeScapes'* in the end.[16] Fostering regional flexibility and quick adaptability to revolving economic demand profiles is, at very least, a big challenge for a nation state which has not yet overcome the bureaucratic inertia of former industrialist education policies.

Another hope for knowledge-based development options to arise in peripheries was nurtured by discussions around the availability of internet communication and tele-work, especially during the 1990s. In the meantime it has become evident that this hope might not be fulfilled, mainly because of the spatial effects of the digital divide, which have not been recognized in public debates up to now. The digital divide has mainly been discussed in terms of the unequal access of different social groups to the internet, to computer-based education and to training in informatics.[17] Residents of economic peripheries were perceived as being disadvantaged because of comparatively small numbers of specialised training facilities, and low degrees of supply with broadband cable nets, etc. There is however, little detailed empirical evidence about central topics here: the intricate dynamics established by the differing digital performance of peripheries, differing local and regional access to facilities, the selective investment in specialised segments of the educational sector, the localisation of different types of knowledge, and the degree of sedentariness,

[15] see above, Chapter 5.1

[16] Concerning the concept of KnowledgeScapes, see Matthiesen (2005) and cf. Matthiesen/Bürkner (2004) on Knowledge Milieus

[17] Haywood (1995), Aichholzer/Schmutzer (2000)

or rather mobility, of the educated workforce. Evidence from an IRS research project on regional knowledge milieus and 'knowledge-based city-regional development' hints at a paradoxical, self-reinforcing trend in eastern Germany's regional development: high unemployment and deficiencies of competence reduce the probability of attracting global players or establishing new knowledge-based branches – even in cases where the relevant enterprises are mobile, and modest as far as locational factors are concerned. In contrast to former waves of out-migration, peripheral regions are now developing a knowledge competency trap, which might turn out to be a severe impediment to interregional competition in the future.[18] [19]

5.4 Conclusion

Is it possible to achieve better spatial cohesion by combating brain drain and the loss of regional competency? And can this be done effectively? Under conditions of accelerating economic peripheralization, the answer to the first question is: hardly, at least not by means of traditional policies of regional 'compensation'. The second question remains thorny and is essentially unresolved. The eastern German experience of upcoming governance paradoxes[19] has shown that singular efforts to improve regional human resource bases merely produce new threats to the regions in question. An investment in education and vocational training without suitable economic embedding structures is likely to inflate out-migration because local people will be enabled to act successfully in more attractive labour markets. In a knowledge-based society there is a critical lower threshold for local competencies and knowledge networks which should not be crossed. Creating isolated (higher) educational structures in peripheral regions in the way preferred by individual eastern German Länder such as Brandenburg seems to hold little promise. It seems to be almost impossible to attract economic 'partners' who might draw benefits from human capital kept as regional 'stock', simply because continuing out-migration prevents this stock from ever reaching a critical mass.

Our first conclusion is therefore a negative one: Creating attractive locations cannot be a task for sectoral education policies alone. Joint, integrated efforts by educational institutions, economic actors and policy-makers are needed in order to develop fitting policies and funding schemes. This implies that more detailed information about the demand for a qualified, specialized workforce at a specific location must be gathered. Recent 'disparitarian' developments within the knowledge economy have encouraged policy makers and governance actors to carefully assess economic developments at the local and regional levels. Parts of this new approach to educational policy are already being implemented in Saxony and Brandenburg, where funding schemes are being re-oriented towards supporting regional growth

[18] Matthiesen (2007)

[19] see above, Chapter 5.1

centres where industrial and service clusters have become large enough to produce a continuous demand for specialised employees.[20] Building training facilities around promising clusters might be an option for the future, albeit many knowledge-based enterprises acquire a suitable, talented workforce at the global level of 'human capital' rather than within the same region. Another problem arises in connection with highly specialized production and services. Since knowledge production according to flexible, interdisciplinary standards[21] faces the problem of a decreasing half-life period of the validity of specialized knowledge, a locally produced workforce might not be able to serve rapidly changing demands for this knowledge, thus forcing entrepreneurs to pass over the region when headhunting for their special purposes.

The recent cluster debate, however, still has one important deficit: It is focused on mere adaptations local and regional actors have made concerning the putatively unalterable requirements of globalization and the expansion of the knowledge economy. *'Fitting into global trends'* seems to be the imperative felt by policy-makers and planners alike. This means that the search for endogenous potentials within the regions in question is oriented towards "globally suitable" assets. From a broad array of potential types of knowledge, only a narrow segment is given validity. Local milieu knowledge and other types of knowledge might also be relevant to the stabilization or even growth of peripheral regions – but up to now they have been politically underestimated.

This recent global adaptation hype has blurred the search for realistic development options within eastern Germany's peripheries. In fact, the majority of regions show rather traditional branch structures, often driven by neo-Fordist production and the limited demand for a highly qualified workforce. Their dependency on conjuncture fluctuations and the changing strategies of corporations and enterprises situated outside the region do not favour them as preferred objects of advanced educational policies. In this situation it is all the more important to have a closer look at branches, regional actors, and inventive micro-networks which are able to develop new economic-cultural-social functions, often within technological recesses. *Spatial pioneers*, as the IRS has called them[22], have already invented the regional level anew in terms of economic functions, culturally embedded production, and tourism, etc. It is worthwhile looking at spatial pioneers even from a more heuristic angle, as divining rods for future regional development, and thinking about furthering their approaches in order to complement or develop alternatives to high-risk adaptation strategies.

[20] Frank (2003); Ministerium für Wirtschaft des Landes Brandenburg (2005), Ribhegge (2005)

[21] cf. the debate on *'mode 2'* of knowledge production, Gibbons et al. (1994), and Nowotny et al. (2001)

[22] Matthiesen (2004)

References

Aichholzer G, Schmutzer R (2000) The Digital Divide in Austria. Vienna

Berg F (2003) Politisch-institutionelle Rahmenbedingungen für die Entwicklung regionaler Handlungsfähigkeit im Land Brandenburg und im Freistaat Sachsen. Berlin (http://www.biss-online.de/download/Thomas_Forum_Ost.pdf, viewed on 12 June 2006)

Berger J (2005) Nimmt die Einkommensungleichheit weltweit zu? Methodische Feinheiten der Ungleichheitsforschung. In: *Leviathan*, Issue 33 (4), pp 464-481

Dienel C (ed.) (2005) Abwanderung, Geburtenrückgang und regionale Entwicklung. Ursachen und Folgen des Bevölkerungsrückgangs in Ostdeutschland. Wiesbaden

Ebendrett H J, Hansen D, Puzicha K J (2002) "Brain Drain" in deutschen Regionen - Effekte von Arbeitslosigkeit und innerdeutscher Migration. Arbeitsberichte: Psychologischer Dienst der Bundeswehr, PSZ PIII 6

Ezcurra R, Gil C P, Rapún M (2003) Inequality and Regional Development within the European Union. European Regional Studies Association Conference Papers, ersa03p333

Gatzweiler H-P, Irmen E (1997) Die Entwicklung der Regionen in der Bundesrepublik Deutschland. In: Friedrichs J (ed.) Die Städte in den 90er Jahren. Demographische, ökonomische und soziale Entwicklungen. Wiesbaden, pp 37-66

Franz P, Rosenfeld M T, Roth D (2002) Was bringt die Wissenschaft für die Wirtschaft in einer Region? Empirische Ergebnisse zu den Nachfrageeffekten und Hypothesen über mögliche Angebotseffekte der Wissenschaftseinrichtungen in der Region Halle. Halle (Halle Institute for Economic Research, Discussion Papers, Issue 163; http://www.iwh-halle.de/d/publik/disc/163.pdf, viewed 6 June 2006)

Gibbons M, Limoges C, Nowotny H, et al. (1994) The New Production of Knowledge. The Dynamics of Science and Research in Contemporary Societies. London, Thousand Oaks, New Delhi

Haase M (2005) 'Brain Drain' in Ostdeutschland? Zum Zusammenhang zwischen Schrumpfungsdynamiken und dem Bildungsniveau der regionalen Nachwuchskräfte. Eine theoriegeleitete Fallstudie in Brandenburg. Berlin

Haywood, Trevor (1995): Info-Rich - Info-Poor: Access and Exchange in the Global Information Society. London

Matthiesen U (2003) Im Sog von Schrumpfungsdynamiken – eine Lernende Region im deutsch-polnischen Grenzgebiet. In: Matthiesen U, Reutter G (ed) (2003) Lernende Region – Mythos oder lebendige Praxis? Gütersloh, pp 89-114

Matthiesen U, Bürkner H-J (2004) Wissensmilieus – Zur sozialen Konstruktion und analytischen Rekonstruktion eines neuen Sozialraum-Typus. In: Matthiesen U (ed.) Stadtregion und Wissen. Analysen und Plädoyers für eine wissensbasierte Stadtpolitik. Wiesbaden, pp 65-89

Matthiesen U (2004) Das Ende der Illusionen – Regionale Entwicklung in Brandenburg und Konsequenzen für einen Aufbruch. In: *perspektive*, Issue 21, pp 97-114

Matthiesen U (2005) KnowledgeScapes – Pleading for a knowledge turn in socio-spatial research. Erkner: IRS-Working paper (available for download at: www.irs-net.de), pp 1-19

Matthiesen U (2007) Wissensmilleus in heterogenen stadtregionalen Räumen Ostdeutschlands – zwischen Innovationsressourcen und kulturellen Abschottungen. In: Koch G, Warneken BJ (eds.) Region – Kultur – Innovation. Wege in die Wissensgesellschaft. Wiesbaden, pp 83-122

Milanovic B (2002) True World Income Distribution, 1988 and 1993. In: *Economic Journal*, Issue 112, pp 51-92

Milanovic B (2005) Half a world: regional inequality in five great federations. Washington D.C. *The World Bank, Policy Research Working Paper Series*, Issue 3699

Ministerium für Wirtschaft des Landes Brandenburg (2005) Branchen-Kompetenzfelder, Branchen-Schwerpunktorte, Regionale Wachstumskerne. Potsdam (available for download at: www.wirtschaft.brandenburg.de/cms/media.php/gsid=lbm1.a.1312.de/Charts_Pressekonferenz.pdf; viewed on 1 April 2005)

Nowotny H, Scott P, Gibbons M (2001) Re-thinking Science: Knowledge and the Public in an Age of Uncertainty. Cambridge

Ribhegge H (2006): Stärken und Schwächen der Neuausrichtung der Wirtschaftsförderung Brandenburg. In: *Berliner Debatte Initial*, Issue 17, (1/2), pp 156-166

Sautter B (2004) Regionale Cluster – Konzept, Analyse und Strategie der Wirtschaftsförderung. In: *Standort*, Issue 4, pp 66-72

SPD Parteivorstand (ed.) (2004) Weimarer Leitlinien Innovation. Unser Land gerecht erneuern. Beschluss des SPD-Parteivorstandes vom 6. Januar 2004 (available for download at: http://www.forscher.de/servlet/PB/show/1483108/2004-01-09-SPD-Weimarer-Leitlinien-01-%5B1%5D.pdf)

6 Environmental Protection needs good Administration: Ecological Applications of the New Legal Instrument "European Groupings of Territorial Co-operation (EGTC)"

Gerold Janssen [1]

Introduction

For decades the European Community's member states have been co-operating with neighbouring states and their subdivisions in border regions, particularly at the regional and local level in correspondence to their respective jurisdiction. This form of neighbourhood co-operation across state borders is today generally termed 'cross-border co-operation'. After a long period in which co-operation in legally non-binding arrangements appeared sufficient, state and administrative practice – in line with the increasing marginalisation of state borders in the course of European integration – now requires some new approaches involving obligatory organisation in transnational relations. A new legal instrument was created for this: European Groupings of Territorial Co-operation (EGTC).[2]

Considering the abundance of possible tasks to which an EGTC could be applied, environmental protection and spatial planning are prominent in view of their inherently transnational character. Some examples of the possibilities offered by this new legal instrument will be described here.

The aim is to abolish the constraints in transnational co-operation which arise from asymmetries in the various legal administration systems. These hinder border regions in establishing effective environmental protection programmes and realising co-ordinated planning.

An EGTC can solve these problems by doing away with auxiliary solutions, such as non-binding agreements. Instead, the EGTC provides a legal entity which can participate in legal relationships. Moreover, all further consequences of formal administrative activity are interconnected (decision-making, management – recruiting/personnel contracts, financial contributions and other financial affairs, liability, independent external audits, etc.)

---

[1] The author wishes to thank the following people for their valuable contributions: Lars Stratmann, *Transnational Strategic Environmental Assessment*; Sylwia Czarnecka-Zawada, *Cross-border Protected Areas*; and Beate Konieczny, *Integrated Coastal Zone Management*. Leibniz Institute of Ecological and Regional Development (IOER), Dresden.

[2] Regulation (EC) No. 1082/2006 of the European Parliament and the Council of 05.07.2006 on a European Grouping of Territorial Co-operation (EGTC), *Official Journal*, L 210, p 19

Various institutions such as the AEBR[3] and the European Parliament[4] have stressed the importance of cross-border, transnational and inter-regional co-operation within wider European integration as well as the importance of introducing a separate objective for territorial co-operation, building on the success of the Interreg initiative. In addition to economic and social cohesion, the European Constitution makes specific mention to territorial cohesion in several places. It is recognised as one of the Union's objectives and shared competences (Article I-3 par. 3, I-14 and II-96). Article III-220 of the Constitution[5] stipulates that particular efforts should be made to support cross-border regions.

Formal and informal cross-border co-operation still suffers from a lack of clarity and corresponding legal difficulties. This is particularly true when the contracting parties are intrastate bodies, such as municipalities, rather than states. In this article current proposals will be presented and discussed, with an analysis of deficits and weaknesses, while potential tasks for an EGTC will be identified.

Future territorial co-operation in Europe will largely encompass cross-border, transnational and interregional collaboration. Spatial co-operation must be promoted by an intensification of co-operation between local and regional authorities in border regions. Member states, as well as regional and local authorities, however, still face substantial difficulties in adapting and applying cross-border, transnational and interregional co-operative measures in the context of various national legislative processes and procedures. Appropriate steps are required to reduce these difficulties. It is essential that cross-border co-operation be improved and that measures necessary to enhance the conditions for the realisation of cross-border co-operation be accepted in order to ensure the harmonious development of the entire Community area and the stabilisation of economic, social and territorial co-operation. One initial step is to examine measures to promote better conditions under which actions of cross-border co-operation are carried out.

This article is divided into five parts. After this introduction, the second chapter (6.1) gives an outline of formal arrangements for cross-border co-operation as well as an explanation of regulations necessary for the creation of an EGTC. The third chapter (6.2) provides an overview of previous experience in formal and material cross-border co-operation and its problems. The fourth chapter (6.3) considers the specific requirements of an EGTC, especially the tasks it has to fulfil, and finally the fifth chapter (6.4) concludes the article with a short summary.

---

[3] Association of European Border Regions (2004), http://www.aebr.net

[4] Resolution on Cross-border and Interregional Co-operation of 16.05.1997 (A4-161/1997, rapporteur: Mr Miller); Resolution on Cross-border and Interregional Co-operation of 09.06.1992 (A3-188/92, rapporteur: Mr Cushnahan); Resolution on Cross-border Co-operation at the Internal Frontiers of the European Community of 12.03.1987 (A2-170/86, rapporteur: Mr Poetschki)

[5] cf. the present Article 158 of the EC Treaty

6.1 Legal Framework

General conventions of cross-border co-operation

The European Outline Convention on Transfrontier Co-operation between Territorial Communities or Authorities Concerning Interterritorial Co-operation from 21 May 1980[6] (Madrid) created a legal basis for intergovernmental conventions, thereby promoting and facilitating co-operation between the regional administrative bodies of the contract states. The intensification and advancement of relations between neighbours is achieved by this so-called umbrella agreement. It is based on two principles: the inviolability of the regional administrative bodies' areas of responsibility, as determined by the states' domestic laws (Article 2), and the priority procedure of domestic law over the final agreement (Article 3, par. 4).

The European Outline Convention was not fashioned as a direct legal basis for transnational co-operation, but requires instruments in the form of further intergovernmental agreements. Such agreements include the BeNeLux Agreement of 1986, the Anholt Agreement of 1991, the Vienna Agreement of 1993 between Italy and Austria, the Rome Agreement of 1993, the Bayonne Agreement of 1993 between France and Spain, the Mainz Agreement of 1996 as well as the Karlsruhe Agreement of 1996. Other conceivable forms of legal arrangement are the creation of Euro districts, the transnational contract sui generis or contracts under private law.[7]

The aim of all the agreements mentioned above is to move from a private law legal basis to a public law basis, under which sovereign tasks can be managed comprehensively and more easily (e.g. environmental protection, spatial planning, etc.).

The Regulation of European Groupings of Territorial Co-operation (EGTC)

The legal situation regarding transnational co-operation between regions and municipalities at the western borders of Europe has been substantially improved by the above-mentioned conventions. At the northern, eastern and southern borders, however, various problems persist. A European legal framework for transnational co-operation is an urgent concern in view of current extensive legal problems and highly diverse statutory arrangements within national laws.

[6] European Outline Convention on Transfrontier Co-operation between Territorial Communities or Authorities (Madrid, 21.V.1980). European Treaty Series (ETS), No. 106. [ON-LINE]. (Strasbourg), Council of Europe, Treaty Office, (15.10.2003). The European Outline Convention on Transfrontier Co-operation between Territorial Communities or Authorities was opened for signature by the Council of Europe's member states and for accession by European states which were not member states on 21 May 1980, and came into force on 22 December 1981.

[7] Kotzur (2006) pp 57f

The legal barriers to cross-border co-operation have been substantially lowered by the creation of a specific and standard European legal instrument. This improves and secures existing agreements as well as regulating and aiding the formation of new co-operation. Following preliminary work carried out by the Association of European Border Regions (AEBR), the European Commission submitted a proposal for a regulation to establish a European Grouping of Territorial Co-operation (EGTC),[8] which came into force on 1 September 2006.

The regulation contains the following points:

- The regulation establishing EGTC allows the creation of co-operative groupings on community territory, invested with legal personality (Recital 8, Article 1, par. 3). The EGTC is given the authority to act on behalf of its members, in particular the regional and local authorities of which it is composed. It may employ staff and be party to legal proceedings (Article 1, par. 4).

- The EGTC may deal with all kinds of 'territorial co-operation', whether cross-border, transnational or inter-regional (see the legal definition of Article 1, par. 2).

- The tasks delegated to the EGTC must be defined by its members in a convention of European territorial co-operation (Article 8 and 9). During the administration of Structural Funds it carries out any co-operative action at the sole initiative of the member states and/or their regional and local authorities (Recital 11, Article 7, par. 3). The tasks and competencies of the EGTC should be set out in a 'Convention' (Recital 10).

- Members of the EGTC can be member states, regional authorities, local authorities and bodies governed by public law (Article 3). Any formation shall be decided at the initiative of its members (Article 4, par. 1).

- Recourse to the EGTC shall be optional (Recital 8). The establishment of an EGTC has to be approved by the member states (Article 4, par. 3).

- The EGTC should establish its statutes and create its own organs, as well as determining rules for the budget and for the exercise of financial responsibility.

[8] Amended proposal for a regulation of the European Parliament and of the Council establishing a European Grouping of Territorial Co-operation (EGTC), 7 March 2004, COM (2006) 94 (final), 2004/0168 (COD).

- The tasks given to an EGTC may not involve the exercise of powers conferred by public law or of duties whose object is to safeguard the general interests of the state or other public authorities, such as the police and regulatory powers, justice and foreign policy (Article 7, par. 4).

- No financial liability shall arise for member states on account of the regulation in relation to an EGTC of which they are not a member (Recital 12, Article 12, par. 3).

- The applicable law for the EGTC should be that of the member state where the EGTC has its registered office (Article 2, par. 1).

- Member states have the right to supervise and control the EGTC (Article 6).

- Organs of an EGTC are the assembly and the director. The assembly shall be composed of representatives of the members of the EGTC (Article 10).

- The EGTC has an annual budget (Article 11).

- The EGTC is published in the Official Journal of the European Union. Upon the publication of this notice the legal capacity of the EGTC shall be recognised in each member state.

The legal basis for the regulation is the third subparagraph of Article 159 of the Treaty, which allows for specific actions to be taken outside the Funds, which are the subject of the first paragraph of that provision, in order to achieve the treaty objective of greater economic and social cohesion.

In order to overcome the obstacles hindering territorial co-operation, the regulation introduces a co-operation instrument at community level which allows for the creation of co-operative groupings in the community territory invested with legal personality. Thus, the EGTC regulation refers only to member states. Although the partners of cross-border co-operation with candidate and third countries are not directly included, they are free to participate (Recital 16).

The regulation was seen as having great political importance. At the first reading the European Parliament voted in favour of the new instrument for co-operation (naturally with several amendments), while the Committee of Regions adopted a resolution strongly supporting the Commission's initiative.[9] Within the member states, some of the regions located along borders are extremely interested in having an instrument which facilitates co-operation with bordering regions in a flexible

[9] Opinion of the Committee of the Regions concerning the Proposal for a Regulation of the European Parliament and of the Council establishing a European Grouping of Cross-border Co-operation (EGCC), *Official Journal*, C 071, 22/03/2005, p 46

way, and which gives them a high degree of autonomy with regard to central administrations and governments. Such an instrument is seen as complementing similar initiatives at both the national level and the level of the Council of Europe (e.g., the Euroregions). The impact of the regulation should not limit the field of application for legal provisions which form the basis of intergovernmental agreements,[10] such as the Karlsruhe Agreement.[11]

In order to restrict the regulation to the minimum, national legislation already regulating equivalent groupings formed under national law shall apply to those issues not covered by the regulation or status of the EGTC.

There was a great deal of discussion as to the possible consequences if the objective of the grouping was not clearly defined, as well as about measures necessary to ensure its legality and the regularity of its acts. This led to some improvements and additions to the text not covered by the original Commission proposal.

First of all, the grouping was renamed the European Grouping of Territorial Co-operation (EGTC) throughout the text, to indicate that all kinds of territorial co-operation may be included: cross-border, transnational or inter-regional. The regulation furthermore excludes any financial liability for member states when the EGTC is applied outside the management of Structural Funds. A major source of difficulty in cross-border co-operation is the different bodies of law in each state. As there is still no international administrative law comparable to the International Private Law, the issue of the validity of laws across borders is unclear. Here the regulation determines that the law of the member state where the EGTC has its office should be applied. Arrangements for supervising EGTC activities must be specified in a clear and transparent fashion, taking into account the fact that the EGTC is composed of bodies belonging to different member states and that the actions of the EGTC as a grouping are based on the legal system of one of the member states and subject to supervision by that state.

As the regulation provides appropriate measures for member states' rights of supervision and control, there is no danger that an EGTC could undermine any state's sovereign power. In addition, any initiative for the creation of an EGTC remains with the potential members, and additional obstacles created by member states are not foreseen.

National laws governing the way in which equivalent groupings are formed should also be applied to EGTC. This can be implemented easily in member states where appropriate standards are already in place (for example in Germany, Article 24, paragraph 1a, Constitutional Act). Other states will first have to improve their legislations in order to provide the necessary legal framework.

---

[10] European Parliament (2005) p 7

[11] An agreement reached between the governments of the Federal Republic of Germany, the Republic of France, the Grand Duchy of Luxembourg and the Swiss Federal Council, acting in the name of the cantons of Solothurn, Basel Stadt, Basel Land, Aargau, and Jura, about international co-operation between territorial entities and local public authorities from 23 January 1996 (BGBl. II in 1997, p 1159).

Invested with a legal personality, the EGTC accordingly executes tasks specifically entrusted to it, taking responsibility and assuming liability for them. The EGTC, as a legal entity, has legal capacity, which is one of the inherent characteristics of legal personality. It may execute the tasks entrusted to it directly (as a grouping), or delegate the execution of those tasks to one of its members.[12] Every entity having a legal personality can establish its own organs in accordance with the law governing its internal operation. It therefore seems advisable that the choice of organisational systems be left up to the EGTC members, subject to the proviso that every EGTC should have an executive committee (i.e. an executive body composed of one or more persons) and an assembly representing all members. It should also be up to EGTC members to specify the arrangements for EGTC representations, such as who should represent the grouping in different fields, and whether it should be represented by one or more persons when concluding agreements, etc.[13]

6.2 Experience and Deficits concerning Cross-Border Co-operation

Considering the serious difficulties member states, in particular their regions and local authorities, must overcome in the implementation and administration of cross-border, transnational, and interregional co-operation under the current legal and administrative conditions, it is clear that suitable measures must be adopted to reduce these difficulties. Reinforcement of cross-border, transnational, and interregional co-operation in the community has also become an urgent task in view of the increase in the number of land and maritime borders following enlargement. Existing instruments have proven to be unsuitable for the organisation of structured co-operation, such as the European Economic Interest Grouping (EEIG)[14] with its economic orientation.[15]

At the Leibniz Institute of Regional and Ecological Development (IOER) different phases of co-operation have been observed against the background of relevant projects for cross-border research. Accordingly, one must differentiate between long-term co-operation over many years in border towns and large co-operation areas. There have been different experiences with co-operation, which is an indication of the different goals and challenges concerning cross-border measures.

[12] European Parliament (2005) p 13

[13] European Parliament (2005) p 18

[14] Council Regulation (EEC) No 2137/85 of 25 July 1985 on the European Economic Interest Grouping (EEIG), Official Journal L 199 of 31 July 1985, corrigenda Official Journal L 247 of 14 September 1985, *Official Journal*, L 124 of 15 May 1990.

[15] New instruments such as Functional Overlapping Competing Jurisdictions (FOCJ) were discussed. cf. Frey/Eichenberger (1999)

In order to overcome obstacles which impair transnational co-operation it is necessary for cities and regions to examine the instruments of this co-operation. In practice, co-operative actions regulated under private law (particularly in the legal form of registered associations) should be replaced where possible by public structures.

Intergovernmental contracts are important in forming a basis for cross-border co-operation. Groupings in cross-border co-operation demand a new, co-operative legal structure, and they form an innovative potential which can aid the European process of integration. The new instrument for the creation of European groupings for territorial co-operation has been heartily welcomed by the Euroregions and the Association of European Border Regions (AEBR). The European border regions desire a standard legal instrument for the creation of public forms of co-operation, in order to become independent of the co-operation forms under private law.

Within the Euroregions one can find numerous examples of excellent results achieved in long-term cross-border co-operation, such as found in the German-Czech Euroregion ELBE/LABE documents.[16] The role and tasks of the Euroregions have to be redefined, however, in view of changing basic conditions. An EGTC would provide an opportunity to combine both parts of the Euroregion in Germany and the Czech Republic under a public legal form and establish a common, jointly run secretariat.[17]

Conclusions from large regions such as the Saxony, Lower Silesia and Northern Bohemia region (the so-called Enlarge-Net-Region) are similar. The question of the need for a common administration for the region was answered with a clear 'yes'. Experience gained from the transnational project ENLARGE-NET led to this vote, in view of the provision of further co-operation.[18]

6.3 Requirements and Tasks for an EGTC

The regulation establishing the European Groupings of Territorial Co-operation (EGTC) aims to provide an appropriate framework for facilitating cross-border, transnational, and interregional co-operation (title, Recital 1 and Article 1). Border regions are often the sites of conflicts which cause additional and specific obstacles to cross-border co-operation. This amendment would give the EGTC the ability

[16] Preußcher (2006) pp 57f
[17] Timm (2006) pp 57ff
[18] ibid. p 57ff

to address these specific obstacles to cross-border co-operation.[19] The tasks and competencies of the EGTC must be specified by its members in a convention of European cross-border co-operation (Article 8).[20]

Integrated Coastal Zone Management (ICZM)

The coastal zone represents ecologically valuable and sensitive areas which are at the same time exposed to ever greater pressures from exploitation. This leads to conflicts of interest between requirements for utilization and the interests of protection. Such conflicts must be defused by provisions co-ordinated and adapted for both the land and sea. At the community level a uniform basis for integrated coastal zone management was developed by the Recommendation 2002/413/EC of 30 May 2002 concerning the implementation of Integrated Coastal Zone Management in Europe,[21] which stresses equal validity for all member states. ICZM is thus understood as a dynamic process which demands co-ordinated and strategic actions at local and regional levels, with an appropriate national framework for orientation and support. In addition, a cross-border approach is recommended in order to further European integration. Based on results of the project Integrated Coastal Zone Management in the Odra-lagoon (ICZM-Odra), which was promoted by the German Ministry of Education and Research, possible tasks for an EGTC are determined in the context of a transnational ICZM between Germany and Poland.

The aim of ICZM is to implement the environmentally sustainable, economically equitable, socially responsible, and culturally sensitive management of coastal zones. The integrity of this important resource should be maintained while allowing for traditional activities and customs that do not threaten sensitive natural areas or endanger the status of the wild species of coastal fauna and flora (Recital 5 Rec. 2002/413/EC). To achieve this aim it is necessary to improve the co-ordination of actions taken by all relevant authorities both at sea and on land in managing the sea-land interaction (Chapter I lit. h). It is therefore important that the management of coastal zones has the support and involvement of relevant administrative bodies at the national, regional and local levels. Two approaches are feasible: The actors either

---

[19] European Parliament (2005) p 12

[20] Under the European territorial co-operation objective, the ERDF shall focus its assistance concerning environmental aspects on the following priorities (Art. 6 lit b REG 1080/2006): water management, energy efficiency, risk prevention, and environmental protection activities with a clear transnational dimension. Actions may include: protection and management of river basins, coastal zones, marine resources, water services and wetlands; fire, drought, and flood prevention; the promotion of maritime security and protection against natural and technological risks; and the protection and enhancement of the natural heritage in support of socio-economic development and sustainable tourism.

[21] A Recommendation of the European Parliament and the Council from 30 May 2002, concerning the implementation of Integrated Coastal Zone Management in Europe, *Official Journal*, L 148, 06/06/2002, p 0024 – 0027.

establish or maintain appropriate links with the aim of an improved co-ordination of the various existing policies, which may be sufficient in certain cases, or they establish a partnership with and between regional and local authorities. When more sectors are involved in the ICZM the necessity of an institutional framework becomes obvious. This partnership can be formed as an EGTC.

Additional problems concerning transnational ICZM lie in co-ordination difficulties between the states. In accordance with chapter V of the Rec. 2002/413/EC, dialogue with neighbouring countries should be promoted in order to compile mechanisms for better co-ordination in the transnational context. A set of legal and institutional factors, however, still exists which obstructs transnational co-operation. These include a lack of administrative and institutional mechanisms and authorities, weak legal stability, overly complicated procedures, weakly developed information systems, insufficient adjustment between administrations and institutions regarding co-operation, as well other issues such as a lack of understanding between participants from different nations.

Against the background of transnational ICZM and the central role of management plans[22] in this, the creation of an accordingly aligned EGTC could contribute to a reduction in the above-mentioned problems. Coastal regions and their authorities are better able to control spatial planning measures as well as improve the general understanding and acceptance of ICZM philosophy within the legal framework of such a public grouping provided by an EGTC. This could lead to a better co-ordination of land and sea actions due to dialogue, and the exchange of information and data in a cross-border manner.

Cross-border protected areas

Stronger co-operation in environmental protection is particularly needed between Germany and Poland. The remedying of transnational environmental impacts (such as air and water contamination) as well as the process of waste removal is inefficient without close cross-border co-operation.

Regional and local authorities are best aware of the problems which exist locally, and they also have the greatest interest in solving them. It is therefore important that they possess the correct instruments with which to find and implement solutions.

Germany and Poland have already concluded numerous contracts for improved intergovernmental co-operation as well as for simplifying international co-operation. These only regulate individual sectors, however, such as water management, disaster control or environmental protection. Thus far no bilateral regulation exists that has a general application, rather than a focus on certain special subjects.[23]

Furthermore, the legal framework governing cross-border co-operation for regional and local administrations is organised differently in the two countries and in their national regulations. For example, the Polish Voiwodships do not have

[22] Janssen (2004) pp 251ff

[23] Bußmann (2005) p 267

the power to conclude an agreement under international law such as possessed by corresponding German municipalities. The units of the Polish municipalities are limited in their possibilities for transnational co-operation. Thus, for example, the development of a so-called transnational 'ad-hoc association' (ad-hoc associations of local authorities, special-purpose associations) is currently impossible in Poland.[24]

An EGTC would establish a basis for co-operation between German and Polish autonomies as well as between public bodies. Current instruments for co-operation are regulated by private law and there is a lack of public competencies for instruments to allow them to act with greater legal force. The existing competencies only grant public bodies consultation rights and rights of participation.

It is interesting to examine how the Euroregion Pomerania would benefit by accepting an EGTC status to help it achieve its tasks. Existing structures would be supplemented by an instrument which provides efficient enforcement without requiring new structures. The scope of duties should include, for example, the transnational protection of nature. The Polish Nature Protection Act[25] includes a provision to create transnationally protected areas in co-ordination with neighbouring countries (Article 2, par. 2). These can be implemented by identifying border areas which have a high natural value and concluding an agreement with the neighbouring country in question. One such protected area already exists: the German-Polish national park 'Lower Odra River Valley' between Hohensaaten (Germany) and Szczecin (Poland) is one example of a cross-border protected area.[26]

The 'Lower Odra Valley National Park' is a natural unit consisting of a Landscape Protection Park in Poland (Article 16 Nature Protection Act) and a National Park in Germany (§ 24 Federal Nature Protection Act). Although the Polish and German sides are subordinate to their own national administrations, they nonetheless co-operate on all issues. An EGTC could assume responsibility for administering the area, including the protected area management according to the Habitats Directive.[27] This directive determines the establishment and conservation of a network of protected areas known as Natura 2000. In these areas the member states are obliged to promote biodiversity by maintaining or restoring certain habitats and species which have a special conservation status. Article 6, par. 4 of the Habitat Directive requires that the overall coherence of Natura 2000 be 'protected'. The directive thus presumes that the 'original network' was coherent. If the derogation regime is used then the situation must be corrected, so that coherence is fully restored. The member states should therefore endeavour to improve the ecological coherence of

[24] Janssen/Czarnecka-Zawada (2005) p 24; Czarnecka-Zawada/Janssen (2005) pp 25 – 33

[25] Nature Protection Act, 16 April 2004 (Dz.U. Nr 92, Poz. 880, Ustawa z dn. 16.04.2004 r. o ochronie przyrody).

[26] Internet: www.nationalpark-unteres-odertal.de.

[27] Council Directive 92/43/EEC of 21 May 1992 on the conservation of natural habitats and of wild fauna and flora, Official Journal L 206 , 22/07/1992, p 7

Natura 2000 by maintaining and developing features of the landscape which are of major importance for wild fauna and flora, as is intended by the German regulation on connecting areas. These features can also be cross-border.

It has to be admitted that without fundamentally homogeneous topographic and topic tables base data on the transnational management of protected areas cannot be effectively achieved. The administrative bodies for protected areas are normally insufficiently aligned in organisational structure and cannot manage transnational co-operation. The proper organisation of appropriate communication and management systems could, however, be arranged with the help of an EGTC.[28]

Transnational Strategic Environmental Assessment (SEA)

The European Directive 2001/42/EC on the assessment of the effects of certain plans and programmes on the environment (SEA-Directive)[29] provides cross-border assessment and participation in accordance with Article 7 of the SEA-Directive. Strategic Environmental Assessments (SEA) must be made in conjunction with regional planning, land use planning on a local level, and technical planning (such as transport planning). If planning is likely to have a significant impact on the environment of a neighbouring state then it must be carried out transnationally.

Transnational SEAs demand and promote a more intensive co-operation of municipalities and regions in border areas. Its methods and procedures are: **(1)** the assessment of cross-border environmental effects on the environmental aspects mentioned in annex I (f), **(2)** cross-border authority and public participation, and **(3)** the cross-border monitoring of environmental effects.

State actors must communicate efficiently to ensure that cross-border assessment and participation can be successfully achieved. They must receive basic information regarding the contents of the planning to be assessed, the legal binding of the planning, as well as the individual contents, the scope and depth of possible planning alternatives, the available data and its quality, legal regulations within the neighbouring state, official structures and competences of the planning and environmental offices on different planning levels, existing monitoring activities, as well as any of the the neighbouring state's further plans and programmes in the cross-border area of the SEA. All transnational assessment and participation activities must solve communication problems resulting from any different languages being used. This particularly affects those neighbouring states with highly diverse languages.

---

[28] Csaplovics/Wagenknecht (2006) pp 57f

[29] Directive 2001/42/EC of the European Parliament and of the Council from 27 June 2001 on the assessment of the effects of certain plans and programmes on the environment, *Official Journal* L 197, 21/07/2001, p 30

Possible tasks for an EGTC in the context of a transnational SEA shall be formulated below on the basis of experience with an INTERREG III A project concerning an SEA for regional planning in the German-Polish-Czech border area and from the perspective of an SEA for land use planning.[30]

In the first appraisal of a transnational SEA in the German-Polish and German-Czech border regions in Saxony, the competences and the tasks of the domestic authorities involved are clear and seem suitable and sufficient for most aspects of the cross-border SEA. It is felt, however, that four aspects of the transnational SEA require additional cross-border coordination and communication, and this can be only be partially achieved with the means and structures presently available. These aspects could be possible tasks for an EGTC within the transnational SEA. They are:

a) The exchange of information on current data and planning bases for the border region, as well as,
b) Mutual information about current plans and programmes in the border region,
c) The coordination and possible execution of cross-border monitoring, also in combination with monitoring for other plans or programmes, where applicable, and
d) Quite simply, a continuous communication flow free of language difficulties during the assessment and participation processes.

Overcoming language barriers represents a central task for the future. Information from neighbouring states regarding their planning and potential environmental effects must often be restricted to only the most essential facts due to the scarce financial resources and a lack of personnel. Proper evaluation of this information is often impossible, however, without further knowledge of details. Additional difficulties arise when the correct translation of technical terms is not given and translations have to be interpreted; the same terms may in fact have a slightly different meaning in neighbouring states. An EGTC could therefore provide communication and co-ordination services for different environmental and planning authorities as well as for administrative institutions in the context of the transnational SEA. It could co-ordinate, translate and communicate detailed information for neighbouring state regions and municipalities on the basis of a broad specialized knowledge of the planning process and procedures, and the legal basis of neighbouring states.

Structural Funds

Many measures of the above-mentioned topics of ICZM, protected areas (Natura 2000), and SEA are partly co-financed by the Structural Funds. The link with this financial instrument will therefore be examined. It is important to note that the regulation of EGTC forms part of a legislative package of new regulations for the

[30] Reinke (2006) pp 133f

Structural Funds during the period 2007-2013, comprising a general regulation to establish a set of common rules which apply to all the instruments and specific regulations covering the European Regional Development Fund (ERDF), the European Social Fund (ESF), and the Cohesion Fund. The new legal instrument for EGTC shall facilitate the implementation of co-operation programmes.[31] It also forms part of a package of internal and external instruments relating to cross-border co-operation inside the European Union (European grouping of cross-border co-operation), with candidate countries or potential candidate countries (pre-accession instrument), and with third countries with which the European Union wishes to establish neighbourly relations (European neighbourhood and partnerships instrument).

The INTERREG programmes have already introduced cross-border projects on a goodwill-basis of common management, financing, guarantees and control. The aim of an EGTC is to increase and intensify the existent transitivity of borders, connecting diverse legal systems and giving regional and local organisations the possibility of co-operating across borders without prior consent from member states. This results in an increase in responsibility.[32]

Although further motivation in favour of creating an EGTC could no doubt be won from the experience gained in current INTERREG III programmes, this is not yet possible, at least in terms of comprehensive and objective estimates. A recent study at the University of Strathclyde concluded that it is still too early to judge the effectiveness of the current INTERREG programmes and their structures. It is therefore not possible to give general recommendations concerning the new funding period and the legal instrument itself. However, some first observations can be made which should be considered in further discussions about the possible employment of European groupings in the framework of structural funds.[33]

Owing to its legal personality (Article 1, par. 3) an EGTC can assume the function of the administrative authority itself, or transfer it to one of its members (Article 7, par. 5). This removes the necessity of burdening national or regional administrations with the rather atypical task of cross-border activity. The legal implications can be ignored since the governing law is still that of the member state in which the EGTC settles as an administrative authority. Nevertheless, one can assume that an EGTC symbolises international co-operation better than a purely national institution. Finally, the regulation presupposes the membership of several member states which will specify and supervise the agenda, including the authority of EGTC organs.[34]

According to the ERDF-Regulation (Article 14, par. 1) a joint technical secretariat must be dedicated to the administrative authority for each programme. The geographical proximity of authority and secretariat is at least suggested if not explicitly determined. Thus it can be argued that from the perspective of

[31] Hübner (2005), Speech/05/557
[32] European Parliament (2005) p 14
[33] Graute (2006) p 111
[34] ibid. p 111

administrative efficiency the authority and secretariat should be located in the same member state, as the same legal framework will automatically determine the administrative actions. Whether this is the most effective approach depends on the co-operation goals. If the aim is to develop a common administrative body operating outside national traditions, it can be of advantage if institutions from different states are merged into the programme management. The members of the grouping must then decide if they wish to co-operate in more appropriate structures or conserve their national frameworks. Neither one of these options should be considered as better or worse. Those who desire particularly good projects will arrange the administrative activity on the programme level as simply as possible. Those who seek the long-term development of common administrative action in a broader sense will probably be interested in a different arrangement and the optimal use of the new legal instrument.[35]

6.4 Conclusions

As stated in the explanatory memorandum and preamble to the discussed Regulation EGTC (Recital 1), the purpose of the regulation is to achieve the objective of social and economic cohesion envisaged by the treaty (Article 158 EC-Treaty) and strengthen cross-border co-operation by improving the implementation conditions for actions of cross-border co-operation. The new legal instrument aims to reduce obstacles and difficulties encountered when managing actions of cross-border, transnational or inter-regional co-operation within the framework of differing national laws and procedures. It does this by creating a co-operation instrument at community level to allow the setting-up of co-operative groupings on community territory, invested with a legal personality.

The results of this survey can be summarised as follows. Due to the progressive development of the acquis communautaire there already exist numerous measures which can be managed by an EGTC. The recommendation of ICZM and the legislation of the Habitats and the SEA Directives are only a few examples of measures which depend greatly on cross-border co-operation. These tasks can be assigned to a transnational grouping such as EGTC.

Integrated Coastal Zone Management takes a holistic approach. More than 13 different sectors and areas are mentioned in Chapter III of the EC Recommendation 2002/413/EC. The strategic approach of ICZM is based on the improved coordination of action taken by all the relevant authorities both at sea and on land in managing sea-land interaction. If the sectoral arrangements are to be overcome, then a new institution possessing the necessary legal powers, such as an EGTC, is required.

[35] Graute (2006) p 111

In the case of spatial planning, municipalities and regions have to co-operate with the neighbouring state regarding the co-ordination and examination of regional and local plans. For each land use plan the accompanying environmental assessment has to be co-ordinated across borders. This can be best achieved by an institution which is familiar with both the environmental data inside its own country as well that of the neighbouring state. Furthermore, the responsible institution should know exactly which requirements the local development plan must fulfil in the neighbouring state. Thus an EGTC is important for transnational trade-offs, the exchange of environmental data, and the production of environmental reports. Once an EGTC has been established, further tasks can also be transferred to it, e.g. all urban land use planning projects. Experience from across the range of cross-border data management has demonstrated that a complete harmonisation does not yet exist. It would be useful to place a database covering all the EGTCs set up on union territory at its disposal, entrusting this task to the Committee of the Regions or the member states.[36] The same applies to the administration of Cross-border Protected Areas.

The transfer of administrative functions to an EGTC can be completed by the transmission of financial autonomy concerning Structural Funds. Numerous projects from the above-mentioned sectors are co-financed by the European Regional Development Fund, the European Social Fund and the Cohesion Fund. If an EGTC is able to act to implement territorial co-operation programmes or projects co-financed by the community, a single system of local or regional development will be the result.

[36] European Parliament (2005) p 17

References

Association of European Border Regions (ed.) (2004) Towards a new Community legal instrument for decentralised public law-based Transeuropean co-operation. Gronau, Internet: www.aebr.net

Bußmann A (2005) Die dezentrale grenzüberschreitende Zusammenarbeit mit Deutschlands Nachbarländern Frankreich und Polen. Baden-Baden

Council Regulation (EEC) No 2137/85 of 25 July 1985 on the European Economic Interest Grouping (EEIG)

Csaplovics E, Wagenknecht S (2006) Grenzüberschreitendes raumbezogenes Management von Schutzgebietsregionen in einem EVTZ. In: Janssen (ed.) (2006) Europäische Verbünde für territoriale Zusammenarbeit. Berlin

Czarnecka-Zawada S, Janssen G (2005) Rechtsinstrumente der grenzüberschreitenden Zusammenarbeit zur Umsetzung eines bilateralen IKZM am Stettiner Haff. In: Glaeser B, Sekœcińska A, Löser N (eds.) (2005) Integrated Coastal Zone Management at the Szczecin Lagoon - Exchange of experiences in the region. *Coastline Reports 6*, ISSN 0928-2734, pp 25 – 33

Directive 2001/42/EC of the European Parliament and of the Council from 27 June 2001 on the assessment of the effects of certain plans and programmes on the environment, *Official Journal*, Issue L 197, 21/07/2001

European Parliament (2005) Report on the proposal for a Regulation of the European Parliament and the Council in establishing a European grouping of cross-border co-operation (EGCC) (COM(2004)0496 – C6-0091/2004 – 2004/0168 (COD), 21 June, Rapporteur: Jan Olbrycht

Council of Europe (2003) European Treaty Series (ETS), No. 106, Strasbourg, Treaty Office, 15.10.2003

Frey B S, Eichenberger R (1999) The New Democratic Federalism for Europe – Functional, Overlapping and Competing Jurisdictions. Edward Elgar Publishing Limited, Cheltenham.

Glaeser B, Sekscinska A (eds.) (2005) Management of the Oder/Odra Coastal Zone, *Coastline Report*, No. 5, Rostock.

Graute U (2006) Strukturfondverwaltung mithilfe Europäischer Verbünde. In: Janssen (ed.) (2006) Europäische Verbünde für territoriale Zusammenarbeit (EVTZ). [European Groupings of Territorial Co-operation (EGTC)] Berlin

Hübner D (2005) Territorial Cooperation and European Objectives, INTERACT conference Steps towards Territorial Co-Operation Programmes 2007-2013, Brussels, 28 September 2005, Speech /05/ 557.

Janssen G (2004) Harmonisation of management plans: Natura 2000, Water Framework Directive and EU Recommendation on ICZM. In: Schernewski G, Löser N (ed.) (2004) Managing the Baltic Sea, *Coastline Reports*, Issue 2, Rostock-Warnemünde, pp 251 f

Janssen G, Czarnecka-Zawada S (2005) Administrative Zusammenarbeit zur Umsetzung eines bilateralen IKZM in der deutsch-polnischen Odermündungsregion. *IKZM-Oder Berichte*, Issue 17

Kotzur M (2006) Rechtsfragen grenzüberschreitender Zusammenarbeit. In: Janssen (ed.) (2006) Europäische Verbünde für territoriale Zusammenarbeit (EVTZ). [European Groupings of Territorial Co-operation (EGTC)] Berlin.

Nature Protection Act (2004) Dz.U. Nr 92, Poz. 880, Ustawa z dn. 16.04.2004 r. o ochronie przyrody, 16 April

Preußcher Ch (2006) Euroregionen – haben sie sich bewährt? – Funktionen, Aufgaben und Perspektiven. In: Janssen (ed.) (2006) Europäische Verbünde für territoriale Zusammenarbeit (EVTZ). [European Groupings of Territorial Co-operation (EGTC)] Berlin

Regulation (EC) No. 1082/2006 of the European Parliament and the Council of 05.07.2006 on a European Grouping of Territorial Co-operation (EGTC), *Official Journal*, Issue L 210

Resolution on Cross-border and Interregional Co-operation of 16.05.1997 (A4-161/1997, rapporteur: Mr Miller)

Resolution on Cross-border and Interregional Co-operation of 09.06.1992 (A3-188/92, rapporteur: Mr Cushnahan)

Resolution on Cross-border Co-operation at the Internal Frontiers of the European Community of 12.03.1987 (A2-170/86, rapporteur: Mr Poetschki)

Reinke M (2006) Grenzüberschreitende Abstimmung in der Strategischen Umweltprüfung. In: Janssen (ed.) (2006) Europäische Verbünde für territoriale Zusammenarbeit (EVTZ). [European Groupings of Territorial Co-operation (EGTC)] Berlin

Timm J (2006) Braucht die "Enlarge-Net-Region" eine eigene Behörde? In: Janssen (ed.) (2006) Europäische Verbünde für territoriale Zusammenarbeit (EVTZ). [European Groupings of Territorial Co-operation (EGTC)] Berlin

7 Suburbanisation, Suburbia and “Zwischenstadt”: Perspectives of Research and Policy

Joachim Burdack, Markus Hesse

Introduction

Theoretic discourses and empirical findings related to the process of suburbanisation and the resulting spatial category of suburbia have long been dominated by critical assessments of metropolitan growth, the decline of inner cities and associated problems (with regard to land use, transport, tax distribution). More recently, following the rising significance of sub- and exurban development in North America as well as in Europe, research and planning attitudes towards suburbia have become more pluralistic. It is now more acknowledged that edge urban developments represent a legitimate component of the urban fabric.

The new research and planning attitudes about urban outskirts are related to the observation of new urban forms, coined in terms like edge city[1], post-modern urbanism[2], post-suburbia[3], or “Zwischenstadt”.[4] Respective discourse points out new qualities in regional development and policy: the spatial category suburbia is no longer only negatively evaluated, but perceived in a differentiated way. It is thus becoming the object of policies and strategies for improvement and further development, rather than that of disregard and negation. In this context, this paper aims at giving a condensed overview of the state and dynamics of the subject, how it is being discussed, and which consequences for research, policy and planning may result from this perspective.[5] This paper provides an overview of the state of suburbanisation research in Germany and Europe and also of the theoretical concepts which are put forward – both in Germany and internationally – to deal with it. The spatial category suburbia which resulted from the dynamic suburbanisation processes in the second half of the 20th century is then analysed. In the foreseeable future, stagnation and shrinking phenomena are likely to shape the agenda again. The paper’s final section therefore discusses which future perspectives exist for suburbanisation and the further development of suburbia in the context of demographic change and urban shrinkage.

---

[1] Garreau (1991)
[2] Dear/Flusty (1998)
[3] Kling et al. (1995)
[4] Sieverts (1997)
[5] For an earlier version of the paper, see Burdack/Hesse (2006)

7.1 Suburbanisation and Suburbia in Germany

In the past few decades urban development in a majority of the highly industrialized countries was characterized by tendencies of spatial de-concentration. This was also true for cities in the Federal Republic of Germany during the period following World War II. The de-concentration process first affected large agglomerations where an out-migration of the population and increased employment created extended suburban zones around the central cities.[6]

Parallel to this process – sometimes with a certain time lag – an increasing de-concentration of economic activities was registered, partly as a reaction to the suburbanisation of the population, as in the case of household-oriented services and partly caused by the intrinsic locational dynamics of certain economic activities, like manufacturing for example.[7] As early as the 1970s space consuming activities like wholesale trade and logistics exhibited a preference for suburban locations with good accessibility.[8] High-level producer services, on the other hand, remained more strongly attached to city centres, with certain exceptions such as the Rhine-Main region and Stuttgart.[9]

Since the 1980s growth dynamics in large agglomerations have been gradually shifting from the old cores to the urban fringes and the rural surroundings.[10] Medium-sized cities outside the metropolitan areas began to form their own suburban rings. Central cities and surrounding areas merged into functional urban regions that now form the spatial basis of daily activity systems for a majority of the population.[11] This process varied in different metropolitan areas, depending on specific historical and spatial settings: monocentric metropolitan areas such as Hamburg or Munich had different spatial patterns than polycentric regions like the Ruhr, Rhine-Main, Rhine-Neckar or Stuttgart, where typical suburban locations had been traditionally mixed with older centres. The Berlin metropolitan area, where the division of Germany had formed two separate territories, presented a special case: For different political reasons, suburbanisation processes took place predominantly within city boundaries until reunification in 1989-90, particularly in the western part of Berlin.

With the expansion of settlement and commuting areas, the system of settlement structures and the central place hierarchy changed as well. The growth of commuting areas often followed the ideal-typical curve of land prices.[12] The more or less economically rational behaviour of actors, who are attracted by low prices for rents and real estate, is generally regarded as a central impetus for suburbanisation. On the supply side, growth strategies of suburban communities with

[6] BBR (2005a) p 191ff
[7] Kahnert (1998)
[8] Hesse (1999)
[9] Eisenreich (2001)
[10] Hesse/Schmitz (1998); Schönert (2003)
[11] Priebs (2004)
[12] Motzkus (2002)

extensive supplies of developable land, which made regional planning controls inefficient, have to be mentioned.[13] While accessibility was an important factor for suburbanisation, the negative effects of high traffic volumes are regarded today as among the most pressing problems of suburban areas. The fact that the once sharp phenomenological distinction between the spatial categories of 'town' and 'country' is increasingly blurred has also been criticised. The adjustment of living conditions, and concomitantly of spatial settlement structures, is, however, an almost inevitable consequence of modernization: The more suburbia appears 'mature', i.e., the higher the settlement densities of suburban locations become, the more heterogeneouse their social structures, and the supplementation of residential uses by other functions becomes more likely. In this context suburban areas begin to resemble the original properties of cities.

The process of reunification in Germany in 1990 represented a big step forward for suburbanisation dynamics in Germany.[14] It especially led to accelerated suburbanisation in eastern Germany, which persisted until the end of the 1990s. Major reasons for this acceleration were a lack of regional planning guidance concerning the limitation of land offers within suburban communities, fiscal incentives for new housing construction, and restrictions on inner city construction due to unsettled claims for property restitution. These factors steered a large portion of the demand for housing and retail facilities to the outskirts. Suburbanisation dynamics have been declining significantly since the end of the 1990s, and came to an almost complete stop in eastern Germany with the exception of the Berlin metropolitan area. Some eastern German urban regions even reported a reversal of the migration direction in favour of central cities.[15] This reversal is likely to be more than just a brief cyclical interruption of a continuous de-concentration tendency. In western Germany the de-concentration process continues, but its focus has shifted from the outer suburban areas to the urban fringes, that is to say, closer to city centres.[16] Counter-urbanization tendencies that were still noticeable in the 1990s have stopped, and the overall intensity of suburbanisation has diminished. Since 2000, large western German cities have once again exhibited positive population development.

Summing up the tendencies outlined above, it can be stated that suburban areas have experienced a substantial – if regionally differentiated – revaluation during the last few decades. They did not separate functionally from the central cities but have become integral parts of newly formed, larger urban regions. The different parts of such urban regions are increasingly differentiated and selectively used in the course of what might be called a *'regionalization of ways of life'*: One lives in the countryside or in the city, depending on income and certain phases of the life cycle, one works either in suburbia or in the inner city, and leisure time occurs both

[13] Aring (1999)
[14] Siedentop et al. (2003); IOER et al. (2005)
[15] Herfert (2002) p 338
[16] Siedentop et al. (2003)

in suburban areas and in metropolitan cultural centres.[17] Thus, the spatial fix-point of the organization of everyday life is no longer the city centre, but the individually shaped networks of activities, which may stretch over the entire urban region and beyond. Urban research and regional planning reacted to these changes by designating new concepts and new spatial categories. The "Raumordnungsbericht 2005" (Federal Report on Spatial Planning and Development) introduced a new spatial category, "Zwischenraum" (intermediate space), which is positioned between the "Zentralraum" (central space) and "Peripherraum" (peripheral space) and which is characterised by specific properties concerning centrality, population potential, and accessibility.[18] Spatial categories for suburban areas are "äußerer Zentralraum" (outer central space) and "Zwischenraum mit Verdichtungsansätzen" (intermediate space with agglomeration tendencies) and are shown in Fig. 1. A different approach by Siedentop[19] defined a radius of 60 kilometres around the centre of an agglomeration as being suburban. By subtracting central cities from the total area inside this circle, it was estimated that about two thirds of the population lived in the suburbs and about half of all employment was located there.[20]

7.2 International Perspectives: Suburbia and Beyond

In a broad perspective, two patterns of suburban development can be differentiated in western Europe.[21] First, urban areas in north-western Europe were dominated by tendencies of suburbanisation in the 1960s and 1970s and afterwards frequently by de-urbanisation, resulting in a population loss within the entire urban region. The large urban regions in southern Europe experienced high population growth in the central cities until the 1970s due to in-migration from rural areas. Second, since the 1980s, a transition to suburbanisation tendencies has been observed, with diminishing population growth rates within central cities, whereas suburban areas have experienced an additional increase. International comparative analyses of suburbanisation processes in Europe pose problems due to the absence of suitable data bases. Attempts to provide updated databases for urban and metropolitan regions in Europe have thus far yielded only limited success.[22] It remains to be seen whether the ESPON project, which identified 1595 functional urban areas and 76 MEGAs (Metropolitan European Growth Areas), will be more successful in this respect.[23]

[17] Priebs (2004)
[18] BBR (2005a); BBR (2005b)
[19] Siedentop et al. (2003)
[20] ibid.
[21] KEG (1991); Rozenblat/Cicille (2003)
[22] e.g. NUREC (1994); GEMACA (1998)
[23] KEG (2004) p 17

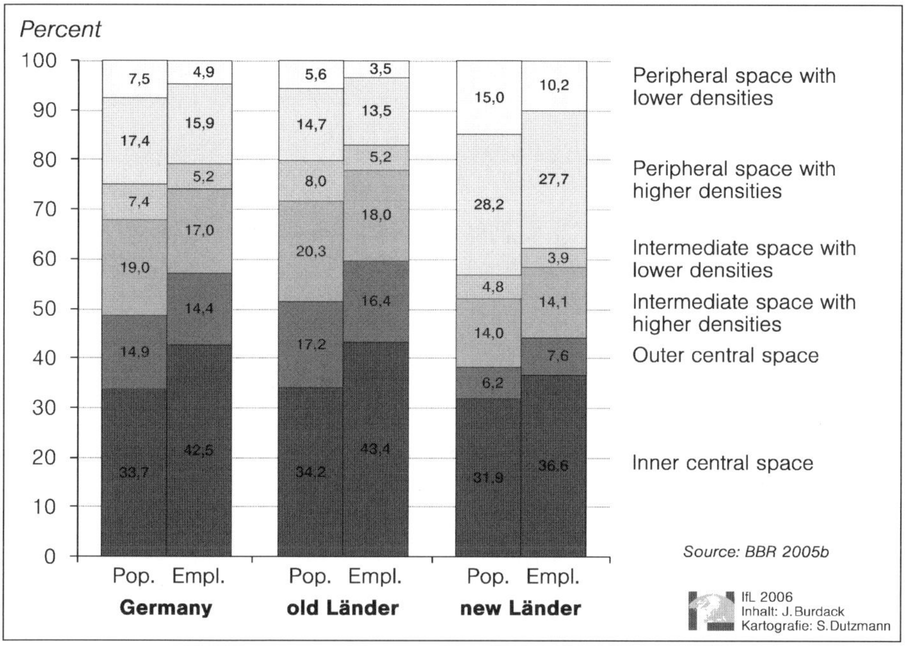

Fig.1: Distribution of Population and Employment according to Spatial Categories 2003
Source: Burdack and Hesse 2006 (modified)

The unsatisfactory data situation is one of the reasons why international comparative analyses are usually performed on the basis of case studies. A study on the outskirts of 11 northern, southern and western European cities,[24] for instance, points out similar spatial structures and development on the urban fringes of all case study areas. The outskirts have developed according to network structures that are based on accessibility, rather than on proximity. Patchwork structures were found to result from the overlay of different networks and the formation of places as nodes within the nets. Residential patterns in the outskirts often follow a principle of maximization concerning contact to green spaces, e.g., ribbons of development or isolated locations. This contributes to the suburban settlement structures having a kind of fragmented appearance.

Suburbanisation research on Central and Eastern European cities effectively started in the 1990s, once socialist planning economies had been replaced by market mechanisms and the first signs of de-centralisation became noticeable. One and a half decades after the demise of socialist systems in Eastern Europe, a considerable body of research exists on suburbanisation processes in post-socialist cities. Much of the work is regionally focused on cities such as Budapest, Moscow, Prague,

[24] Dubois-Taine (2003); Borsdorf/Zembri (2004)

Warsaw or other capital regions, where such processes have been most visible.[25] Despite the regional differences, some general conclusions may be drawn from this research:

- Residential suburbanisation of middle-class households has become an important trend. The new suburban settlements do not form complete new rings around the central city, but are concentrated in certain geographic sectors or along certain transport corridors. Large parts of the urban peripheries are inhabited by lower-class populations. Social disparities between upper-class new suburbanites and the lower-class, old-established population are prevalent.
- Large scale suburban shopping centres form an important part of the metropolitan retail trade, for instance in Èerni Most and Zlièín (Prague) or Budakalász (Budapest). International retail chains were the principal investors (e.g. Auchan, Metro, Tesco). They were searching for accessible sites along highway interchanges or subway terminal stations. Logistic centres and modern manufacturing sites are also increasingly found on metropolitan peripheries. There are only a few examples of more complex centres or new economic poles that also include offices and leisure facilities. The poles of Budaörs-Törökbalint southwest of Budapest and Chimki, northwest of Moscow, should be mentioned here. Due to the 'over-bounding' of many Eastern European cities (cities which have boundaries that extend far beyond their developed areas), typical suburban development can often be found within the urban fringes inside the administrative limits of the central city.

Studies of urban areas in Central and Eastern Europe have frequently focused on the question of whether suburbanisation processes within the transformation countries follow Western European development paths or not. Instead of a simple 'catch up' type of development, 'hybrid' development patterns of urban peripheries seem to emerge instead.[26] These developments consist of different mixes of **(1)** persisting structures developed under socialist conditions, **(2)** elements of transition and transformation, and **(3)** new suburban and post-suburban spatial structures.

There is also significant evidence that the outskirts in many European metropolitan regions have entered a phase of re-concentration. To an increasing extent, such 'post-suburban' multi-functional spaces[27] show characteristics which had long been limited to analyses made of city centres.[28] Economic poles developed in the

[25] e.g., Andrle (2001); Brade/Nefjodowa (1998); Burdack et al. (2004); Degórska (2003); Fassmann/Matznetter (2005); Jakóbczyk-Gryszkiewicz (2002); Kok/Kovacs (1999); Ladányi/Szelényi (1999); Ouředníček (2005); Rudolph/Brade (2003); Rudolph/Brade (2005); Sykora (1999); Sykora et al. (2000); Tammaru (2001); Timar/Varadi (2001); Węcławowicz (2005); Welch/Guerra (2001)

[26] Burdack et al. (2005)

[27] Kling et al. (1995)

[28] Burdack/Herfert (1998); Phelps/Parsons (2003)

outskirts are sometimes complementary to those found within central cities, but may compete with them as well. A well-researched example is the city of Zurich, where a complex patchwork of different land uses has been developed in Zürich-Nord. Not only branch offices are found here, but also national and international headquarters of multinational companies.[29] Another example is the London borough of Croydon, which was transformed from a dormitory suburb into a significant office and retail centre in the 1960s and 1970s. In more recent years, however, Croydon's growth has stagnated due to the rapid ageing of its office supply and increasing congestion problems.[30]

Many concepts have meanwhile been developed to describe this phenomenon of new nodes or economic poles outside central cities. The best known term regarding North American development is 'edge city'[31], with alternatives including 'technoburb'[32] and 'exopolis'.[33] Such terms, however, often refer to one specific type of new centre outside the traditional city centre and specifically refer to the North American situation.[34] Ruth Rohr-Zänker[35] considers it unlikely that edge cities will ever emerge in Western Europe. Radical changes in the settlement structure in Europe will be held back by the different cultural values associated with cities, differences in the planning systems, and the contrasting form of political regulation. Higher population densities and denser urban networks mean that medium-sized towns in Western Europe often act as the focus, around which new functional centres in the outer hinterland of the metropolises crystallise. However, the economic conditions that favour the formation of new activity clusters outside central cities do exist in parts of Western Europe as well. Both in North America and Western Europe, the shift from industrial to post-industrial societies produced a shift of agglomeration advantages from the city scale to the scale of the urban region.[36]

A study of new nodes and activity clusters in the outskirts of continental European metropolitan areas pointed at the emergence of polycentric structures.[37] New economic poles of edge city-like dimensions were found to have emerged in most of the regions studied. Prominent examples of new economic poles include the Büdaörs-Törökbálint area west of Budapest with its proliferation of commercial activities and offices[38] and the Tres Cantos new town north of Madrid with an R&D and high tech-manufacturing profile[39]. The emergence of new polycentric structures

---

29 Hitz et al. (1992)
30 Phelps (1998)
31 Garreau (1991); Jonas (1999)
32 Fishman (1987)
33 Soja (1992)
34 Dear/Flusty (1998)
35 Rohr-Zänker (1996)
36 Phelps/Ozawa (2003)
37 Burdack et al. (2005)
38 Izsák/Probáld (2003)
39 Burdack (2002)

was found to be especially relevant in Paris (see Fig. 2). The Paris metropolitan region and the Dutch Randstad could be seen as two of the most advanced regions in Europe in terms of a more polycentric regional structure, in which new economic centres emerge as competitors to, and complementary partners of, the central cities. It might not be accidental that post-suburban development is most advanced in the two metropolitan areas that are located in the core region of the European space economy, the global integration zone (GIZ).[40] This fact points to a link between the dynamics of spatial restructuring and the level of integration into the global economy.

Generally speaking, the development of European metropolitan peripheries is, however, less dynamic than discourses on 'postmodern urbanism'[41] have ever suggested. European metropolitan areas clearly do not follow a trajectory comparable to that of Los Angeles. There is no decentring of the centre taking place in major European metropolitan areas. Rather, a pronounced intra-regional division of labour between the inner cities and metropolitan peripheries is maintained. High-level producer services and head-office functions remain prevalent for the most part to the inner cities, while the metropolitan peripheries in Europe are gaining strength as locations of research and development and higher education.[42] The changes that are currently taking place do not reveal signs of a 'dissolution' of European cities, but may rather be described as a re-scaling of urban activities on the spatial scale of the urban region.

7.3 Main Concepts of Suburbanisation Research

Suburbanisation research in Germany reached a first peak during the 1970s, parallel to the accelerated development of suburban areas.[43] It is somewhat surprising that studies from that time had been concentrating on a relatively limited set of issues: Important topics were **(1)** the determinants of processes leading to different intensities of suburbanisation, **(2)** the analyses of sub-processes such as the suburbanisation of population, services, manufacturing and trade, as well as **(3)** the repercussions of suburbanisation on the inner cities, which were generally seen as problematic, and **(4)** the effects of suburbanisation on the rural surroundings, which were perceived both in the sense of a revaluation of these areas and critically as "Zersiedelung" (urban sprawl).

New themes and approaches revealed a reviewed interest in questions of suburbanisation in the 1990s. First of all, sectoral aspects of suburbanisation found more interest, that is to say non-residential functions like transportation, leisure

[40] According to Mehlby (2000)
[41] Dear/Flusty (1998)
[42] Bontje/Burdack (2005)
[43] ARL (1975)

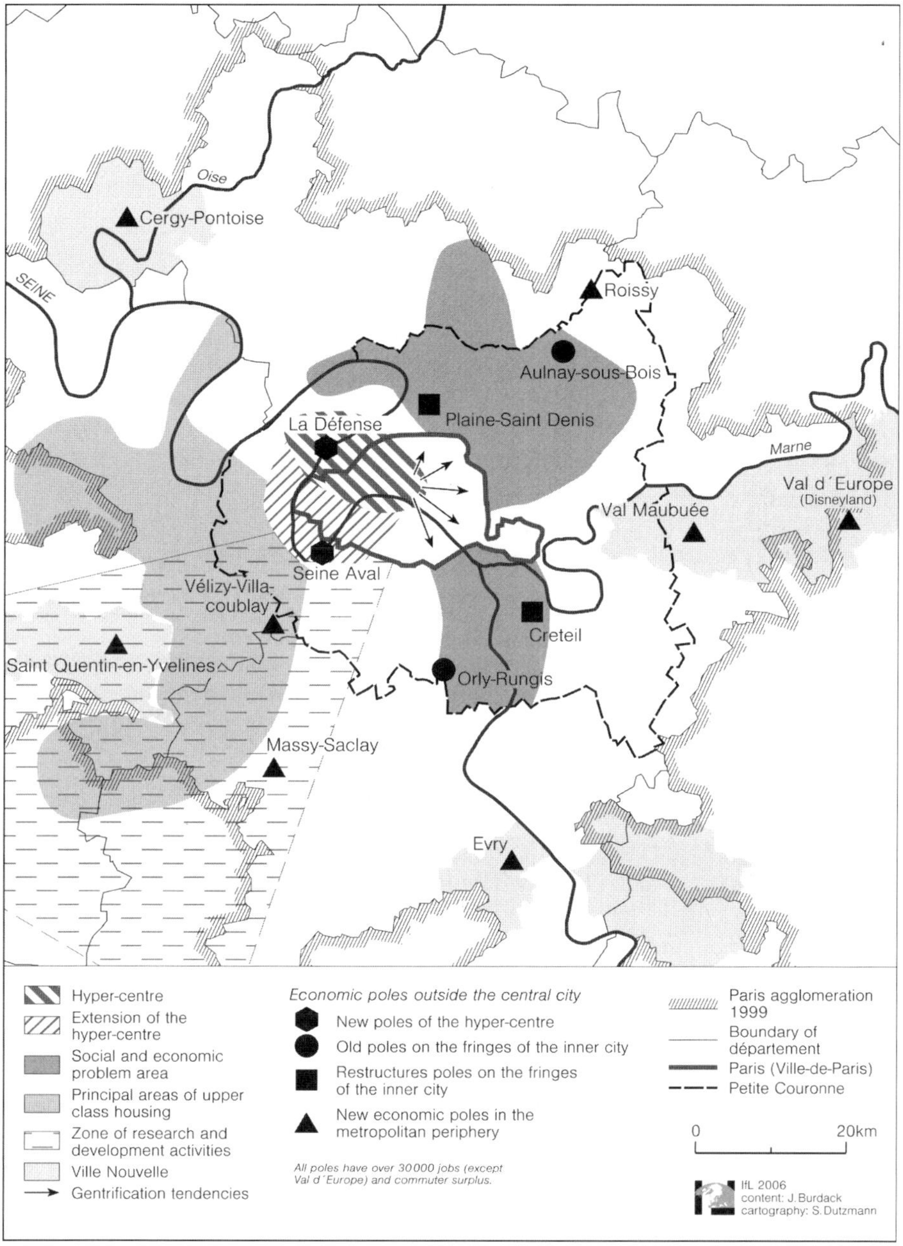

Fig.2: Social and economic polarisations in the Paris agglomeration. Source: Burdack (2004) modified.

activities, trade and services gained wider attention.[44] Secondly, actors on the micro-level and their motivation were analysed in numerous migration studies.[45] Thirdly, influenced by a discourse on the post-modern city, a change in perspective took place that revised the former critical judgements on suburbanisation to a certain degree: The spatial category 'suburbia' was no longer only critically evaluated, but regarded in a differentiated fashion. Such changes in perspective were, particularly in the German context, related to the concept of *"Zwischenstadt"* (in-between city).

Despite a substantial body of research, the term suburbanisation is still not clearly defined and is employed in a variety of fashions in German geographical research. Many topics of suburbanisation research in Germany are filed under different headings in other scientific cultures. A distinction, for instance, is made between *suburbanisation* and *urban sprawl* in British research. The term 'suburban area' designates single family and/or semi-detached housing areas in the outer parts of a city that were mostly constructed during the pre-war and post-war periods.[46] Urban sprawl on the other hand refers above all to forms of 'gluttonous' land use, monotonous development, bad traffic routes and infrastructure, as well as environmental damage and a lack of open space.[47] In French research, the dispersed patterns of settlement growth which occurred during the last few decades are addressed as *périurbanisation*[48].

The wide usage of the 'suburbanisation' label in German research somewhat blurs its content. This can be illustrated by the example of the term *"suburbaner Raum"* (suburban space). The term is used in at least three different ways. Frequently, the surrounding municipalities of a central city are simply called "suburbaner Raum." In this respect, the term is used according to the metaphor of space as a 'container'.[49] In this context "suburbaner Raum" begins at a city's boundaries, regardless of whether the actual land use patterns and spatial structures differ on either side of this dividing line or not. Crucial for labelling an element of the physical-material world as 'suburban' is only its location within a marked area on the surface of the earth.

A second use identifies "suburbaner Raum" as an intermediate spatial category based on density values, the properties of which are positioned between those of urban and rural areas. "Suburbaner Raum" thus becomes a cipher for dispersed settlement structure. In this context the urban area can extend beyond the border of a central city. This is, for instance, the case in the delineation of the city as settlement (morphological city). As outlined above, the new spatial classification scheme of

[44] Brake et al. (2001)
[45] IMU-Institut für Medienforschung (2001); Matthiesen (2002); Blotevogel/Jeschke (2003)
[46] Harris/Larkham (1999)
[47] Peiser (2001)
[48] Dézert/Metton/Steinberg (1991)
[49] Weichhart (1998) p 78

German regional planning policy[50] distinguishes between "Zentralraum" (central space), "Zwischenraum" (intermediate space), and "Peripherraum" (peripheral space) on the basis of the characteristics of density, population potential, and accessibility.

A third concept understands "suburbaner Raum" in the sense of 'suburbia' as a development phase or a 'settlement layer' on the urban periphery. Suburbia is the result of specific spatial processes, particularly the migration of young middle class households to the urban periphery and the related relocation of household services and retail activities that were made possible due to the wide availability of private cars. This more 'relational' use of the term "suburbaner Raum" forms the basis for sociological studies on the 'suburban way of life'.[51] In this context 'suburbia' may be interpreted as a certain phase in a broader development pattern. The concept of 'post-suburbia' which is particularly discussed in the US-American context[52] is an attempt to describe the urban periphery's next phase of development, following suburbanisation as we have known it until recently.

7.4 Suburbia in the Planner's Discourse: Polarized Interpretations and Blind Spots in the Discourse

For a long time the discussion about suburbanisation in Germany was dominated by a critical assessment. This assessment was based on allegedly objective findings concerning the costs and benefits of de-concentration and dispersion on the one hand, and on more subjective estimates concerning suburban settlements' lack of urbanity and poor architectural quality. These evaluations emerged from implicit or explicit comparisons of suburban settlements with inner city locations, particularly concerning settlement densities, urbanity, infrastructure costs and traffic generation. It is certainly true that individual decisions about the location of households and firms caused substantial costs and negative externalities, particularly in respect to infrastructure, traffic, and the environment. Thus, there are considerable societal costs associated with suburbanisation that have not been taken into consideration to their full extent. The simple disqualification of suburban areas as anti-urban and unsustainable, however, overlooks the fact that the criteria for that kind of evaluation are mainly derived from characteristics of inner cities, and are thus not very useful. Traditional terms and concepts do not always contribute to the explanation or solution of new problems. Peter Wilson referred to the juxtaposition of 'good' historical city and 'ugly' suburban patch-work structures as 'propagandistic polarities'.[53]

---

[50] BBR (2005)

[51] Gans (1968)

[52] Kling et al. (1995)

[53] Wilson (1995) p 15

Thomas Sieverts' contribution[54] to the current suburbanisation discourse in Germany, mainly associated with the term "Zwischenstadt" (in-between city), fits into this line of argument. Sieverts demands that suburban developments should be discussed without prejudice as a new type of decentralised settlement structure. The term "Zwischenstadt" hints at Rowe's 'Middle Landscape'.[55] The topology of the "Zwischenstadt" was developed from its intermediate character, which results from characteristics of the built environment as well as from cultural dispositions. The concept of the "Zwischenstadt" was widely discussed in professional and academic circles in Germany. In Germany at least, no other single issue concerning urban development and planning in the 1990s triggered a comparable debate and dispute. The "Zwischenstadt" marked a blind spot in theoretical discourses on urban planning and development. At a more analytical level, however, the "Zwischenstadt" left open many questions. It remained unclear what exactly was meant by this term, which in fact relates to a deliberately 'fuzzy' use of the concept. The term, in fact, has different interpretations: It addresses **(1)** the classical suburban areas at the outskirts of agglomerations, but also **(2)** such parts of suburbia that lie in between different central cities and exhibit rather hybrid settlement characters, and finally **(3)** rural areas where densification and urbanization tendencies are beginning to take place, and which have usually been referred to as the 'urban periphery'. This particular discourse thus left behind substantial problems of definition, which can only partially be accounted for by the variety of suburban settlement structures. Another issue is to what degree Sieverts' arguments can be generalised: The concept of the "Zwischenstadt" was developed on the basis of the Ruhr District and the Rhine-Main Region, two prototypical polycentric metropolitan regions which conform much less to the classical image of a city with its surrounding suburbs than most other urban regions in Germany do. The "Zwischenstadt" studies were continued between 2002 and 2005 in the framework of the *"Ladenburger Kolleg"*.[56] The research efforts of the "Ladenburger Kolleg" were concentrated on the Rhine-Main Region, a metropolitan region where all the elements of the "Zwischenstadt" are present: Successive processes of growth and dispersion, differentiation, and re-concentration had formed a mosaic of old village centres, new housing subdivisions and old industrial sites, which contradicts all the traditional images of the European city. The different components of the "Zwischenstadt" filled up the open areas between the old town centres, structured them and made them 'central' in very specific ways. Some sub-areas profited from their proximity to the old town centres. Behind this almost irritating multitude of development there still lingers the above-mentioned problem of definition: The space of the "Zwischenstadt" consists of old industrial cities (Hoechst, Rüsselsheim) located next to new, service-oriented locations (Eschborn), as well as many fast spreading residential locations, industrial parks, and shopping facilities – thus very heterogeneous land uses, to

[54] Sieverts (1997); Sieverts (2003)
[55] Rowe (1991)
[56] Boelling/Sieverts (2004)

which there is hardly any common denominator. How long these development paths will continue to function also remains unclear. Official regional planning policy in Germany finally recognized the secular trend of suburbanisation and is now going to acknowledge suburban areas as a legitimate spatial category.[57] Yet, a precise definition of suburban areas that takes their diverse characteristics into account is still missing, despite a growing number of research activities devoted to 'suburbia' and also an intensified discourse about the question of how to deal with it in terms of policy and planning.

7.5 Perspectives on the Future of Suburbanisation Processes

The most recent discussion on urban development in Germany has taken place against the background of changing empirical trends: Since the turn of the century the suburbanisation process has come to a stop in many regions of eastern Germany and partly in western Germany as well. In most eastern German regions the growth processes subsided by the end of the 1990s: Out-migration and demographic change now increasingly dominate population and regional development. It is not yet foreseeable whether this turn-around is merely a pendulum swing or a secular turn, that is to say, the 'long goodbye' of an urban expansion process based on growth and suburbanisation.

Demographic and economic growth that translated into residential expansion is often considered fundamental to the suburbanisation process. An answer to the question of to what extent this settlement type can persist under the conditions of declining population figures and a stagnating economy depends on three factors, whose parameters are difficult to estimate at present: **(1)** the slope of the demographic curve, **(2)** the regional distribution of the population, and finally **(3)** the specific space requirements of households and firms, which in turn depend on different social, cultural and economic factors.

According to the Eleventh Co-ordinated Population Forecast of the "Statistisches Bundesamt", the German population will decline until the year 2050 from 82.4 million at the end of 2005 towards a total ranging between 69 and 74 million by 2050, depending on different scenarios including different reproduction and migration rates. Until 2030 the degree of change appears to be limited, whereas the decrease during the following two decades is predicted to accelerate significantly.[58] Given that there are some uncertainties with such long-term predictions, this development might have noticeable consequences for particular regions: Current estimates foresee the development of suburban growth belts around the central cities of northwest Germany, the Rhine axis between Bonn and Mannheim, as well as in many parts of Bavaria.[59] The outskirts of the Berlin metropolitan area and

[57] BBR (2004)
[58] Statistisches Bundesamt (2006)
[59] BBR (2006)

those of several large cities in Saxony will continue to gain population as well. The old-industrialized regions of the Ruhr District and eastern Germany, as well as peripheral rural areas, are, however, particularly threatened by demographic decline. Economic and demographic shrinking and welfare losses will be especially concentrated in the "Neue Länder" (new federal states).

And yet, we do not expect that tendencies of suburbanisation will stop under the conditions of a shrinking and stagnation of urban development, or that problems associated with suburbanisation will disappear automatically. Many of the factors and determinants that supported the suburbanisation process continue to function. The decreasing settlement pressure associated with shrinking has not yet provided any sufficient evidence about an end of suburbanisation, because suburbanisation is not only a reaction to high settlement pressures in the central city, but is also caused by urbanisation dis-economies, e.g., traffic congestion, environmental pollution, and a perception of rising crime levels. The significance of such motivating factors has already been expressed as "Stadtflucht" (flight from the city) by Heuer and Schäfer.[60] This escape-metaphor can also be found in the US-American context as 'flight to the suburbs', as well as in estimates of a decline of suburbanisation in Germany.[61] A significant increase in re-urbanisation would require that central cities become the primary search areas of suburban households and enterprises, which is surely not the case in a general sense. It should be added that, given the functional change inner cities have undergone in becoming locations for high level service centres, new offices, and centres for leisure, tourism, and other events, a return to the city of, for instance, large-scale industrial premises or shopping centres, hardly makes sense .

Nuissl and Rink[62] even argue that the containment of suburban development under the conditions of shrinking could become rather more difficult than in the context of growth. In the course of shrinking, an intensified competition among municipalities for investors and inhabitants is to be expected. Potential investors may have a strong bargaining position and can play municipalities off against each other. The informal, consensus-oriented instruments of German regional planning policy (e.g., "Regionalkonferenzen") are based on 'win-win' situations and have not yet been tested in respective 'win-loose' situations.

In the longer run we do indeed assume that a tendency towards the development of fragmented, perforated urban landscapes will become more likely. The old central city will be a part of this urban region, just as the different forms of "Zwischenstadt" are. The different parts of an urban region might even develop more or less precarious forms of coexistence, depending upon the specific conditions of growth and decline, and stretching or contraction of the urban spaces. The different centres and peripheries will probably be involved in a more intense competition with

[60] Heuer/Schäfer (1978)
[61] Blotevogel/Jeschke (2003)
[62] Nuissl/Rink (2004) p 35

one another. This process is in conflict with our traditional image of the city, but is essentially merely a continuation of urbanisation processes that started in the industrial age.

After the acceleration of the suburbanisation process in Germany in the 1990s, a regionally differentiated process of urban restructuring can be expected in the near future. This restructuring will affect both inner cities and urban peripheries. Even if urban shrinking should contribute to the further stagnation of suburbanisation (as a process), 'suburbia' or "Zwischenstadt" as a spatial category will remain in existence and not disappear. This also means that this spatial category must be considered in planning efforts, either by an improvement in existing structures and an 'orderly retreat' in the context of stagnation and shrinking, or in the sense of prospective planning in growing urban regions. An adequate characterisation of suburbia is still missing in the German discussion. Many labels attached to suburban developments contain explicit criticisms of unwanted developments, like "Amerikanisierung," "Zersiedelung," or 'urban sprawl'. They are not suitable to describe suburbanisation in Europe or to predict uncertain future patterns of development.[63] According to conventional thinking and related discourse in policy and planning, the general goal of reducing the consumption of space appears to be undisputed. However, complaints about 'urban sprawl' or "Zersiedelung" tend to ignore the fact that the issue is much more complex than is often suggested. Critical assessments of suburbanization, for instance, are often determined by urban boundaries, thus criticizing development beyond city limits, and superseding similar impacts within the boundary. Also, suburbia as the spatial frame is made responsible for certain effects that are mainly derived from individual perception and behaviour. Finally, suburban and exurban settlements represent a significant part of the metropolitan landscape, so they should indeed no longer be ignored. It appears that the periodically hegemonic discourse on "sprawl" is an outcome of social construction, rather than being based on material evidence and convincing arguments.[64] Consequently, if suburbanisation is to be better understood than before, it also requires the development of a diversified inventory of terms which takes into consideration both the changing, non-linear process of suburbanisation and its differentiated product 'suburbia'.

[63] Brueckner (2000); Peiser (2001)
[64] Hesse/Kaltenbrunner (2005)

References

Andrle A (2001) Suburbanisierung der Hauptstadt Prag und der Stadt Brünn in den 90er Jahren. In: Akademie für Raumforschung und Landesplanung (eds.) (2001) Erscheinungen, Wirkungen und Steuerungsmöglichkeiten von Suburbanisierungsprozessen. Ergebnisse eines internationalen Planerforums in Prag. Hannover, Issue 276, pp 123-131

Aring J (1999) Suburbia – Postsuburbia – Zwischenstadt. Hannover

Akademie für Raumforschung und Landesplanung (ARL) (eds.) (1975) Beiträge zum Problem der Suburbanisierung. Hannover, *Forschungs- und Sitzungsberichte*, Issue 102

Bundesamt für Bauwesen und Raumordnung (BBR) (eds.) (2004) Suburbia. *Forum Bau + Raum*, Issue 1, Bonn

Bundesamt für Bauwesen und Raumordnung (BBR) (eds.) (2005a) Raumordnungsbericht 2005, Berichte, Issue 21, Bonn

Bundesamt für Bauwesen und Raumordnung (BBR) (2005b) Arbeitspapier Raumstrukturtypen. Konzept – Ergebnisse – Anwendungsmöglichkeiten – Perspektiven. Bonn

Bundesamt für Bauwesen und Raumordnung (BBR) (2006) Raumordnungsprognose bis 2020. Berichte, Issue 23, Bonn

Blotevogel HH, Jeschke M (2003) Stadt-Umland-Wanderung im Ruhrgebiet. Abschlussbericht zum Forschungsvorhaben. Universität Duisburg-Essen, Duisburg

Bölling L, Sieverts T (2004) Mitten am Rand. Auf dem Weg von der Vorstadt über die Zwischenstadt zur regionalen Stadtlandschaft. Wuppertal

Bontje M, Burdack J (2005) Edge Cities European-style: Examples from Paris and the Randstad. In: *Cities*, Issue 22 (4), pp 317-330

Borsdorf A, Zembri P (eds.) (2004) European City Structures. Insights on Outskirts. Brüssel

Brade I, Nefjodowa J (1998) Entwicklungstendenzen und Perspektiven der Stadt-Umland-Prozesse in Russland. In: Europa Regional 6 (4), pp 23-34

Brake K, Dangschat J, Herfert G (eds.) (2001) Suburbanisierung in Deutschland. Aktuelle Tendenzen. Opladen

Brueckner J (2000) Urban Sprawl: Diagnosis and Remedies. In: *International Regional Science Review*, Issue 23 (2), pp 160-171

Burdack J (2002) Massy-Saclay, Tres Cantos, Chimki-Selenograd: Neue Wachstumsräume am Rande europäischer Metropolen. In: Jahrbuch Stadterneuerung 2002. pp 77-91

Burdack J (2004) Die Ville de Lumières und ihre Schatten. Wirtschafts- und sozialräumliche Differenzierungen in der Pariser Metropolregion. In: *Geographische Rundschau*, Issue 56 (4), pp 32-39

Burdack J, Dövényi Z, Kovács J (2004) Am Rand von Budapest – Die metropolitane Peripherie zwischen nachholender Entwicklung und eigenem Weg. In: *Petermanns Geographische Mitteilungen*, Issue 148 (3), pp 30-39

Burdack J, Herfert G (1998) Neue Entwicklungen an der Peripherie europäischer Großstädte: ein Überblick. In: *Europa Regional*, Issue 6 (2), pp 26-44

Burdack J, Herfert G, Rudolph R (eds.) (2005) Europäische metropolitane Peripherien. In: *Beiträge zur Regionalen Geographie Europas*, Issue 61, Leipzig

Burdack J, Hesse M (2006) Reife, Stagnation oder Wende? Perspektiven zu Suburbanisierung, Post-Suburbia und Zwischenstadt: Ein Überblick zum Stand der Forschung. In: *Berichte zur deutschen Landeskunde*, Issue 80 (4), pp 381-399

Clapson M (2003) Suburban Century. Social change and urban growth in England and the USA. Oxford/New York

Dear M, Flusty S (1998) Postmodern Urbanism. In: *Annals of the Association of American Geographers*, Issue 88 (1), pp 50-72

Degórska B (2003) Impact of urban sprawl on environmental planning – Warsaw metropolitan area in the context of European experiences. In: *Europa*, Issue XXI (10), pp 51-60

Dézert B, Metton A, Steinberg J (1991) La périurbanisation en France. Paris

Dubois-Taine G (eds.) (2003) Outskirts of European Cities. State of the Art Report. Cost Action C 10. Brüssel

Eisenreich D (2001) Standortwahl und wirtschaftliche Verflechtungen unternehmensorientierter Dienstleistungsbetriebe in der Filderregion. Tendenzen der Abkopplung suburbaner Räume. Europäische Hochschulschriften V 2706, Frankfurt/ Main

Fassmann H, Matznetter W (2005) Stadtentwicklung in Ostmitteleuropa: Konvergenzen, Divergenzen, Transformation. In: *Geographische Rundschau*, Issue 57 (10), pp 52-59

Fishman R (1987) Bourgeois utopias – The rise and fall of suburbia. New York

Frey W (2003) Melting Pot Suburbs. A Study of Suburban Diversity. In: Katz B, Lang R W (eds.) Redefining Urban and Suburban America. Washington D.C., pp 155-179

Gans H J (1968) Urbanism and suburbanism as ways of life. In: Pahl R E (eds.) Readings in Urban Sociology. Oxford, pp 95-118

Garreau J (1991) Edge City – Live on the new frontier. New York

Group for European Metropolitan Area Comparative Analysis (GEMACA) (1996) North-West European Metropolitan Regions. Geographical boundaries and economic structures. Paris

Harris R, Larkham P (1999) Suburban Foundation, Form and Function. In: Harris R, Larkham P (eds.) Changing Suburbs: Foundation, Form and Function. London, pp 1-31

Herfert G (2002) Disurbanisierung und Reurbanisierung. Polarisierte Raumentwicklung in der ostdeutschen Schrumpfungslandschaft. In: *Raumforschung und Raumordnung*, Issue 60 (5-6), pp 334-344

Hesse M (1999) Der Strukturwandel von Warenwirtschaft und Logistik und seine Bedeutung für die Stadtentwicklung. In: *Geographische Zeitschrift*, Issue 87 (3-4), pp 223-237

Hesse M, Kaltenbrunner R (2005) Zerrbild Zersiedlung. Anmerkungen zum Gebrauch und zur Dekonstruktion eines Begriffs. In: DISP 160, pp 16-22

Hesse M, Schmitz S (1998) Stadtentwicklung im Zeichen von "Auflösung" und Nachhaltigkeit. In: *Informationen zur Raumentwicklung*, Issue 7/8, pp 435-453

Heuer H, Schäfer R (1978) Stadtflucht. Stuttgart.

Hitz H, Schmid C, Wolff R (1992) Zur Dialektik der Metropole: Headquarter Economy und urbane Bewegung. In: *Geographische Zeitschrift*, Issue 80, pp 57-83

IMU-Institut für Medienforschung (2002) Raus aus der Stadt? Untersuchung der Motive für Fortzüge aus München in das Umland 1998-2002. München

IOER, IRS and Universität Leipzig (eds) (2005) Mobilität im suburbanen Raum. Neue verkehrliche und raumordnerische Implikationen des räumlichen Strukturwandels. Forschungsbericht für das BMVBW (FE 70.716). Dresden, Erkner and Leipzig, unpublished

Izsák E, Probáld F (2003) Le développement de l'agglomeration de Budapest: l'exemple de Budaörs. In: *Revue géographique de l'Est*, Issue 12 (1/2), pp 61-68

Jakóbczyk-Gryszkiewicz J (2002) Model of development of suburban zones. The example of major Polish agglomerations. In: *European Spatial Research and Policy*, Issue 9, pp 41-51

Jonas A E G (1999) Making Edge City: Post-suburban development and life on the frontier in Southern California. In: Harris R, Larkham P (eds) Changing suburbs: Foundation, Form and Function. London, pp 202-221

Kahnert R (1998) Wirtschaftsentwicklung, Sub- und Desurbanisierung. In: *Informationen zur Raumentwicklung*, Issue 7/8, pp 509-520

Kommission der Europäischen Gemeinschaften (KEG) (1991) Europa 2000. Perspektiven der künftigen Raumordnung der Gemeinschaft. Luxemburg

Kommission der Europäischen Gemeinschaften (KEG) (2004) Territorialer Zusammenhalt Zwischenbericht. Luxemburg

Kling R, Olin S, Poster M (eds) (1995) Postsuburban California: The Transformation of Postwar Orange County. Los Angeles

Kok H, Kovacs Z (1999) The Process of Suburbanisation in the Agglomeration of Budapest. Netherlands Journal of Housing and the Built Environment 14 (2), pp 119-141

Ladányi J, Szelényi I (1999) Sozialräumliche Polarisierung und Suburbanisierung in Ungarn. In: *Zeitschrift für Wirtschaftsgeographie*, Issue 43 (1), pp 1-15

Lowe M (2000) Britain's Regional Shopping Centres: New Urban Forms? In: *Urban Studies*, Issue 37, pp 261-274

Matthiesen U (ed.) (2002) An den Rändern der deutschen Hauptstadt. Opladen

Mehlbye P (2000) Global Integration Zones – Neighbouring Metropolitan Regions in Metropolitan Clusters. In: *Informationen zur Raumentwicklung*, Issue 11/12, pp 755 - 762

Motzkus A (2002) Dezentrale Konzentration – Leitbild für eine Region der kurzen Wege? Auf der Suche nach einer verkehrssparsamen Siedlungsstruktur als Beitrag für eine nachhaltige Gestaltung des Mobilitätsgeschehens in der Metropolregion Rhein-Main. In: *Bonner Geographische Arbeiten*, Issue 107, Sankt Augustin

Nuissl H, Rink D (2004) Sprawl und Schrumpfung: Das Beispiel Leipzig. In: Nuissl H, Rink D (eds) Schrumpfung und Urban Sprawl. Analytische und planerische Problemstellungen. UFZ-Diskussionspapiere, Department Ökonomie, Soziologie und Recht, Issue 3/2004, Leipzig, pp 24-67

Network of Urban Research in the European Union (NUREC) (1994) Atlas of Agglomerations in the European Union. Duisburg

Ouředníček M, (2005) New suburban development in the Post-socialist city: the case of Prague. In: Eckhardt F (eds) Paths of urban Transformation, Frankfurt/M., pp 143 - 156

Peiser R (2001) Decomposing urban sprawl. In: *Town and Planning Review*, Issue 72 (3), pp 275 - 298

Phelps NA (1998) On the edge of something big. Edge-city economic development in Croydon South London. In: *Town Planning Review*, Issue 69 (4), pp 44 - 465

Phelps NA, Parsons N (2003) Edge Urban Geographies: Notes from the Margins of Europe's Capital Cities. In: *Urban Studies*, Issue 40 (9), pp 1725 - 1749

Phelps NA, Ozawa T (2003) Contrasts in agglomeration: proto-industrial, industrial and post-industrial forms compared. In: *Progress in Human Geography*, Issue 27 (5), pp 583 - 604

Priebs A (2004) Vom Stadt-Umland-Gegensatz zur vernetzten Stadtregion. In: Jahrbuch StadtRegion 2003. Opladen, pp 17 - 42

Rohr-Zänker R (1996) Neue Zentrenstrukturen in den USA. Eine Perspektive für dezentrale Konzentration in Deutschland? In: *Archiv für Kommunalwissenschaft*, Issue 35 (2), pp 196 -225

Rowe P (1991) Making a Middle Landscape. Cambridge

Rozenblat C, Cicille P (2003) Les villes - européennes: analyse comparative. DATAR - La Documentation Française, Paris

Rudolph R, Brade I (2003) Die Moskauer Peripherie – postsowjetische Entwicklungspfade und globale Integration. In: *Osteuropa*, Issue 9, pp 1400 - 1415

Rudolph R, Brade I (2005) Moscow: Processes of restructuring in the post-Soviet metropolitan periphery. In: *Cities*, Issue 22 (2), pp 135-150

Schönert M (2003) 20 Jahre Suburbanisierung der Bevölkerung. In: *Raumforschung und Raumordnung*, Issue 61 (6), pp 457-471

Siedentop S, Kausch S, Einig K, Gössel J (2003) Siedlungsstrukturelle Veränderungen im Umland der Agglomerationsräume. *Forschungen*, Issue 114, Bonn

Sieverts T (1997) Zwischenstadt. Braunschweig/Wiesbaden

Sieverts T (2003) Cities without Cities. An Interpretation of the Zwischenstadt. London, New York

Soja E (1992) Inside exopolis – Scenes from Orange County. In: M. Sorkin (ed.) Variations on a theme park. The new American city and the end of public space, New York, pp 94-122

Statistisches Bundesamt (ed.) (2006) Bevölkerung Deutschlands bis 2050. 11. Koordinierte Bevölkerungsvorausberechnung. Wiesbaden

Sýkora L (1999) Changes in the Internal Spatial Structure of Post-communist Prague. In: *GeoJournal*, Issue 49 (1), pp 79-89

Sýkora L, Kamenický J, Hauptmann P (2000) Changes in the spatial structure of Prague and Brno in the 1990s. In: *Acta Universitatis Carolinae – Geographica*, Issue 35, pp 61-76

Tammaru T (2001) Suburban Growth and Suburbanisation under Central Planning: The Case of Soviet Estonia. In: *Urban Studies*, Issue 38 (8), pp 1341-1357

Timar J, Varadi MM (2001) The Uneven Development of Suburbanization during Transition in Hungary. In: *European Urban and Regional Studies*, Issue 8 (4), pp 349-360

Węcławowicz G (2005) The Warsaw metropolitan Area on the Eve of Poland`s integration into the European Union. In: Hamilton I, Dimitrovska-Andrews K, Pichler-Milanović N (eds) Transformation of cities in Central and Eastern Europe – towards globalization. Tokyo, pp 223-247

Weichhart P (1998) "Raum" versus "Räumlichkeit" – ein Plädoyer für eine transaktionistische Weltsicht der Sozialgeographie. In: Heinritz G, Helbrecht I (eds) Sozialgeographie und Soziologie. *Münchener Geographische Hefte*, Issue 78, Passau, pp 75-88

Welch Guerra M (2001) Ein neuer Typ der Suburbanisierung. Siedlungsstrukturen und Stadtplanung in der tschechischen Metropole Prag. In: *RaumPlanung*, Issue 94, pp 26-30

Wilson P (1995) Euro-Landschaft. In: Westfälischer Kunstverein (eds) Die verstädterte Landschaft. Ein Symposium. München, pp 13-23

8 Varying Costs of Living as Compensation for Regional Disparities in Germany

Rupert Kawka

Introduction

Regional disparities in material wellbeing can be used as one possible indicator of the extent to which the goal of equivalent living conditions, as postulated by the *"Grundgesetz"* art. 72 and *"Raumordnungsgesetz"* par. 1, has been achieved. The statistical variable "average disposable income per capita" is the best measurement for this topic. Looking at this variable, strong regional disparities in Germany become apparent. For example, a representative person living in the counties Hochtaunuskreis near Frankfurt (Main) or Starnberg near Munich has – when looking at these statistics – a standard of living which is twice as high as that of an inhabitant of Weimar and the Kyffhäuserkreis in Thuringia or of Uecker-Randow, Ostvorpommern, or Demmin (all in Mecklenburg-Western Pomerania): the first group had a disposable per capita income of 26,900 to 27,600 Euro; the latter one's ranged between 13,000 and 13,500 Euro in 2004 (cf. Fig. 1, p 103).

But it is an overhasty interpretation to take only the disposable income as an indicator of material wealth and a standard of living, as it merely shows the nominal income. Many goods and services are also more expensive in regions with higher incomes, or, generally speaking, the costs of living (COL) in such areas are higher.[1] Real income should therefore be used instead. The example of rents may illustrate this (cf. Table 1, p 102).

Rents are important in this respect, because they make up about one fifth to one forth of consumer expenditures, making them the largest share of these expenditures (cf. Table 2, p 102). Table 1 shows that rents increase with nominal disposable income, and in the end, if all income would be used only for rents, more or less the same floor space could be rented in every German region.[2]

But in reality, of course, not all income is used for rents. The question about how the other 75% to 80% of the basket of commodities (i.e. all other goods and services except rents) affects nominal income has to be answered as well. Do they also rise, on balance, in the same way as nominal income; are the other prices more or less the same all over Germany's regions or – which is hard to imagine – are they

[1] The Federal Constitutional Court ("Bundesverfassungsgericht") has ruled that civil servants in regions with high costs of living cannot demand a bonus to compensate for high prices, but it also says that legislators have the duty of monitoring the development of the COL in the regions (BVerfG, 2 BvR 556/04 of 6.3.2007, Absatz-Nr. 71, http://www.bverfg.de/entscheidungen/rs20070306_2bvr055604.html).

[2] Kawka (2006a) p 358

Table 1: Disposable income and rents. Source: www.vgrdl.de, BBR Housing Market Monitoring System, author's own calculations.

Disposable income per capita in Euro 2004	< 15,000	15,000-17,500	17,500-20,000	> 20,000
Average rent per m² in Euro 2004	4.83	5.02	5.93	7.05
No. of administrative counties	99	176	123	41

Table 2: Share of expenditures in % of sections of the basket of commodities 2003. Source: Statistisches Bundesamt (2004a), p 72 and pp 82f.

Section	In %		
	Germany (average)	Two regional examples	
		Hamburg	Thuringia
Food, tobacco	14	14.2	15.4
Clothing, footwear	5	5.4	4.8
Rents	24.1	25.4	21.7
Energy	5.7	5.2	6
Furniture, household appliances	5.7	5.3	6
Health care	3.7	3.6	2.6
Transport	14.4	10.6	15.2
Communication	3.1	3.2	3.1
Recreation, culture	11.8	13.1	13.1
Education	0.9	1.4	0.8
Hotels, restaurants	4.3	5.5	4

lower in high-income regions? Only an examination of all the sections of the basket of commodities – or at least of a very high share – will allow for a statement to be made about whether real income is more or less the same all over Germany; whether there is an overcompensation – i.e., that regions with a lower nominal income have a higher real income – or whether regions with a higher nominal income also have a higher real income, so preserving the rank order of the regions.[3] Up to now, however, such a conclusion cannot be drawn, as the necessary information is lacking. This article will nevertheless try to answer this question as well as possible.

[3] Focusing on high and low nominal income does not imply that it should be regarded as the only influence on the COL. Further factors can be population density, the integration of a region in the market, i.e. the distance to the next supplier, the mobility of the inhabitants of a region, regional scarceness, the share of tradeable goods in the basket of commodities (here, for example, the costs of waste disposal, local taxes or administrative fees etc. can be subsumed, i.e. goods and services where the provider has a regional monopoly), regional differences in the preferences of consumers and, of course, regional variations in the quality of goods and services.

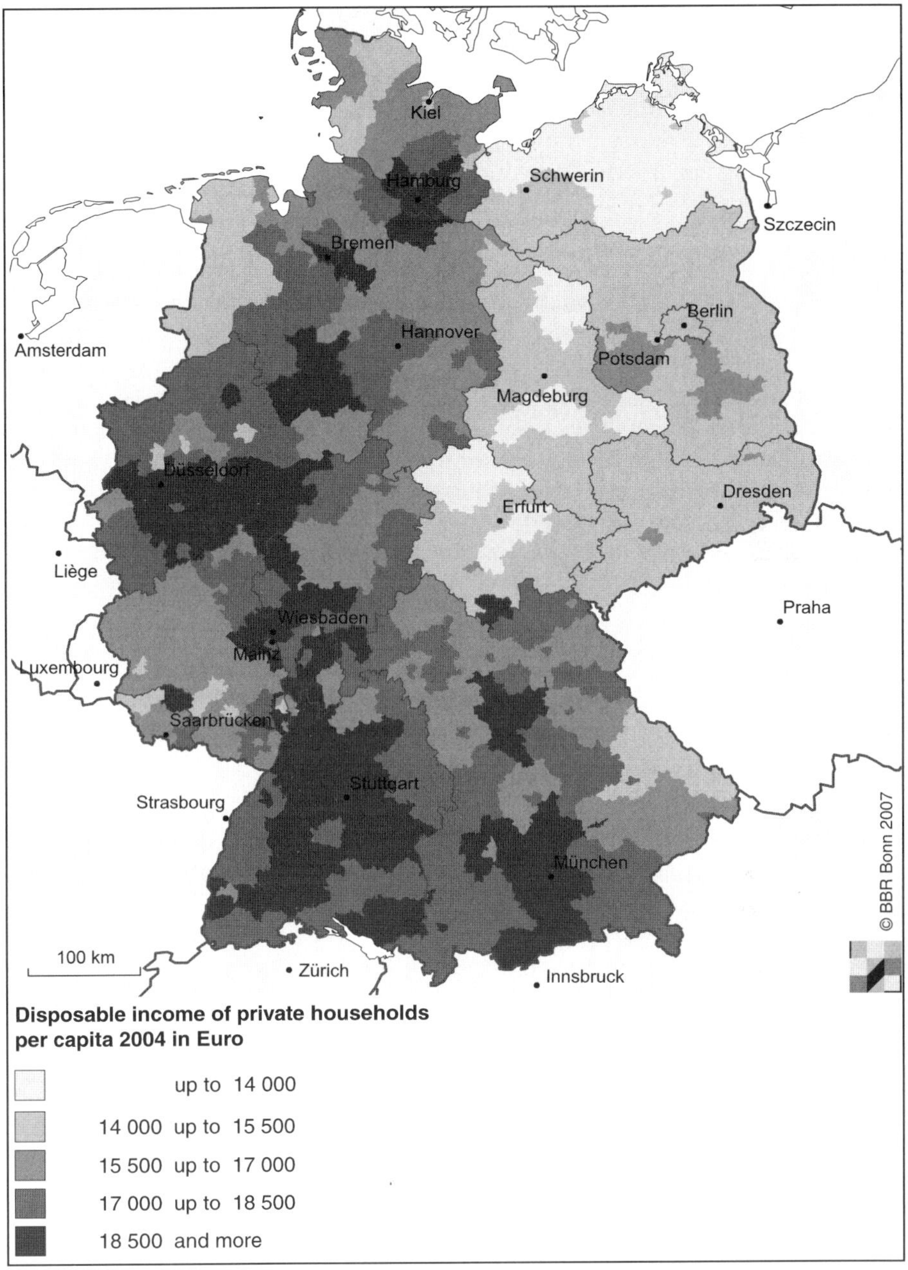

Fig.1: Disposable per capita income 2004. "Kreise" (counties) as of 31.12.2004. Source: Arbeitskreis "Volkswirtschaftliche Gesamtrechnungen der Länder".

Information on regional COL[4] is of great interest. For spatial planning it is important to know whether and to what extent market forces, i.e., the reaction of prices to regional settings (e.g., high-versus low-income regions, urban versus rural places, central versus peripheral locations, etc.), help to achieve the goal of equivalent living conditions, as mentioned above. But this is not the only reason. Knowledge about the COL and real income is important for migration studies[5] and for employers, as they may adjust wages to attract new employees to high-cost areas like Munich. Additionally, much literature about the number of persons living in poverty or with low incomes exists when for example, COL in urban versus rural regions are taken into account.[6] The discussion about the adjustment of unemployment benefits ("Arbeitslosengeld II") in the old and the new "Länder" (federal states) in 2005 needs to be looked at as well in this context. COL should be taken into account concerning questions about income convergence within or between countries as well.[7] Furthermore, there is the need for transparency concerning the charging of fees by local or regional authorities. Citizens have to pay very different prices for the same or at least for comparable services. In the communities of North Rhine Westphalia, for example, fees for waste disposal ranged between 109 and 937 Euro in 2005 for comparable households[8], and the costs for a given quantity of waste water ranged between 246 and 1,321 Euro in the same year.[9] Area-wide studies could help to find out regions or regional administrative bodies with good and bad price performance ratios.[10]

8.1 Methodological Problems

Surveying COL and collecting data in different regions involves coping with six problems in particular. These difficulties are not new; they have been discussed for the past few decades.[11]

- Differences in the quality of goods and services cause the biggest obstacle, as they influence the price, too. Few goods, such as fuel or electricity, are homogeneous, but most items in the basket of commodities are hard to compare in a regional

[4] It would be more correct to speak of a spatial price index.
[5] cf. e.g. Cullision (1984) pp 27ff
[6] cf. e.g. Siminski/Saunders (2003); Curran et al. (2005); Jolliffe (2006)
[7] cf. e.g. Persson (1994)
[8] cf. http://www.steuerzahler-nrw.de/download/abfall2005neu7.pdf
[9] cf. http://www.steuerzahler-nrw.de/download/abwasser2005.pdf
[10] Some regional administrative bodies use the differences in these fees to promote their locations as being attractive for residential purposes due to low COL (cf. e.g. Kreisverwaltung Ahrweiler (1999) p 6).
[11] Evers/Schmuckler/Teich (1963); Kohlhuber (1992); Fenwick/O'Donogue (2003); Statistisches Bundesamt (2004b)

perspective (e.g., theatre tickets, fees for kindergartens, etc.). Differences in quality can be partially levelled. A sophisticated approach is the use of hedonic price models. Here, additional information about the goods is collected and their influence is determined with the help of a regression model.[12] A less complex way is to look for goods and services which can be – at least to some extent – standardized. One example is an English course offered as an evening class for adults ("Volkshochschule"), lasting either 45 or 60 minutes. The same applies to a Pizza Margherita. Most pizza services offer it with a diameter of 26 cm, but some also sell 28 cm pizzas. The didactic competence of the teacher or the taste of the meal can still differ, but the influence of the duration of the lesson or the size of the pizza can somehow be neutralised. Both examples, although nearly trivial, show that the problem of differences in quality can be handled to some extent. But a great number of other items remain which are hard to compare.

- There is an ongoing discussion about whether identical or regionally adjusted baskets of commodities are more suitable. The advantage of the first approach is that the index for COL is comparable throughout the regions. This is not possible while going the second way, as weighting for the goods and services varies (cf. Tab. 2, p 102). The latter method, however, is regarded as more realistic. Consumers, for example, have to pay a higher share of their expenditures for housing in urban than in rural areas, but it is the other way around concerning the costs for transport. Thus, the weights are endogenous and reflect consumers' reactions to regional settings.

- Differences in prices within regions have to be neglected – a common assumption in regional science. But this raises the question of whether the level of the counties or that of the planning regions ("Raumordnungsregionen", an aggregation of four to five counties with a major town as the centre) is more appropriate for such an analysis. The larger the regions, the greater the chance that all items in the basket of commodities can be bought there, but also the stronger the obliteration of intraregional disparities.

- The strong assumption has to be made that consumers purchase all goods and services in their places of living. This neglects reality, because it is not possible to buy all items in the basket of commodities in rural regions, for example. And the inhabitants of suburbs, which often belong administratively to another district than the neighbouring town, regularly visit this town for shopping due to there being more on offer in cities than in the suburban villages. Somehow, distorted results have to be expected.

[12] cf. e.g. Kokski/Cardiff/Moulton (1994); Engels (2003); Chowhan/Prud'Homme (2004)

- It is impossible to collect price data without having a high number of surveyors directly in the regions. For example, if only one single price in every region had to be collected for each item in the section food and non-alcoholic beverages in the basket of commodities, more than 61,000 prices (155 items à 394 "Kreisregionen") would be necessary. Chain stores may offer a trick to sidestep this problem, as they very often have the same prices all over Germany. Furthermore, the goods they offer are of comparable quality. But this approach raises two other problems: It has to be assumed that the consumer will only buy the products there, which is a strong simplification. Furthermore, the question arises of how to cope with the regions when there is no shop belonging to this specific chain store.

- With limited manpower, the internet is the main source of price-related data. Direct surveys in the regions are not possible, so there will be a time lag between the collection of the information and the presentation of the results.

Thus, the surveying of COL will always be an abstraction, as many assumptions have to be made. At best, it is possible to attain information about COL for a representative consumer with an average income and an average expenditure structure. Therefore, the aim of such a study is to approximate reality as well as possible. But, as other countries like Great Britain, Japan and the USA regularly make these interregional comparisons,[13] the difficulties discussed above do not seem to be unsolvable, or can be sidestepped to some extent at least. Furthermore, many of the enumerated problems also exist in international comparisons which provide highly accepted results in research and politics (e.g., to achieve equivalent real incomes for EU civil servants in all member states).[14]

8.2 A new Glance at existing Data

In Germany there are three main existing surveys which have collected original data on COL. The first one dates back to 1978.[15] The study investigated 31 towns and cities in the western part of Germany which had populations ranging from 45,000 to 1,910,000 inhabitants. A second analysis, as in 1979 also carried out by the Federal Statistical Office, was performed in 1993[16] – taking both West and East German towns and cities into account which had between 43,000 and 2,176,000 inhabitants. The major weakness of the two studies is the disregard of one important item in

[13] cf. Wingfield/Fenwick/Smith (2005); Ministry of Internal Affairs and Communications (2005) p. 559; cf. Internet: www.coli.org

[14] For a discussion about the methodological difficulties with the topic in Great Britain see Ball/Fenwick (2003).

[15] Rostin (1979)

[16] Ströhl (1994)

the basket of commodities, namely rents. At that time, however, it was not possible to collect adequate data on rents for standardized apartments. The third study was carried out by the Bayerisches Staatsministerium für Wirtschaft, Verkehr und Technologie (Bavarian Ministry of Economic Affairs, Infrastructure, Transport and Technology) in 2003, using data from 2002.[17] This study provides information about 21 Bavarian communities which have between 12,000 to 1,235,000 inhabitants. This survey thus contains information about smaller communities as well. Information about rents is included, too. But the study only analysed 37.3 % of total expenditures according to the basket of commodities for the whole sample.

In addition to these three very important publications, some studies about Germany or parts thereof also exist. They either look at small samples or focus on no more than a few components in the basket of commodities.[18]

So far, no area-wide surveys about COL in Germany exist – neither on the level of the 439 districts or aggregates of the districts, like "Kreisregionen" or "Raumordnungsregionen", nor for one of the 16 federal states. Michael Roos provides an estimate based on the 1993 data,[19] but his methodological approach is not entirely free of problems. For example, he even transfers information collected in urban places to rural areas. Furthermore, he takes the size of the population as an exogenous variable in his regression model, but this method will be discussed at a later point.

The following paragraphs mainly extract some new information from three studies made by Werner Rostin (1979), Gerd Ströhl (1994) and the Bayerisches Staatsministerium für Wirtschaft, Verkehr und Technologie (2003), as they themselves are predominantly descriptive in character. The differences between them could be regarded as an obstacle to drawing some universal conclusions. This can also be seen as a strength too, because results which point to the same conclusion but which are based on different spatial-temporal samples can be assumed to be comparatively robust. The results are as follows:

First, the use of nominal income as a statistical variable for the material well-being of the inhabitants of a region is misleading, as areas with low nominal incomes can have comparatively high real incomes due to low COL or vice versa. Speaking of nominal income, Munich's was higher than anyone else's in 2002, but by taking COL into account, 13 out of the 20 communities in the Bavarian sample (65%) had higher real incomes than Munich.

---

[17] A kind of predecessor to this study is Gesellschaft für Konsumgüterforschung (1982), which has data for Bavaria as well.

[18] e.g. Struff/Wilamowitz-Moellendorff (1983); Angermann (1989); Buschle (1997); Mehnert (1997); Bund der Steuerzahler NRW (2005a, b) for North Rhine Westphalia, Bund der Steuerzahler Niedersachsen/Bremen (2005) for Lower Saxony, http://www.bissev.de/Wasserpreise/Wasserpreistabelle.xls for Saxony and www.steuerzahler-rheinland-pfalz.de/files/204/Abfall_Tabellen.pdf for Rhineland-Palatinate

[19] Roos (2003) pp 15/21

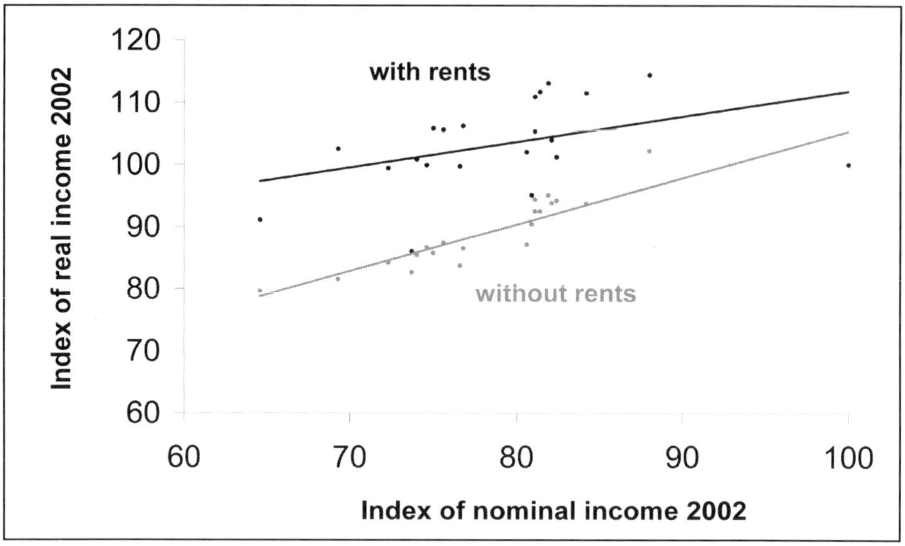

Fig.2: Relationship between nominal and real income with and without rents in Bavaria, 2002, Sources: Bayerisches Staatsministerium für Wirtschaft, Verkehr und Technologie (2003) p 38, author's own calculations.

Both cases in Fig. 2 – either with rents or without them – show an ascending trend line and, thus, a positive relationship between nominal and real income. An interpretation of the slopes shows that especially COL including rents balances out regional disparities in nominal income. Although quite a high number of Bavarian regions have a lower nominal and a higher real income than Munich, it is not possible to conclude that the poorest regions tend to be the wealthiest in the end. Furthermore, the question of whether the slopes are significantly positive has to be asked. A simple linear OLS regression model can show that real income with rents is not dependent on nominal income, i.e., that the prices compensate for the differences in nominal income.[20] But this result is attained only when Munich is part of the sample – without taking the Bavarian capital into account, nominal income still has a significantly positive impact on real income. Other analyses tend to support this latter and modified hypothesis. For example, studies on the COL in 46 towns in Queensland, Australia[21], in 29 Bavarian communities[22], in ten regions

[20] But the probability is only slightly above the 5% percent significance level of rejecting the null hypothesis.

[21] cf. Queensland Office of Economic and Statistical Research Queensland/Office of Economic and Statistical Research, Queensland Treasury (2001)

[22] Gesellschaft für Konsumgüterforschung (1982)

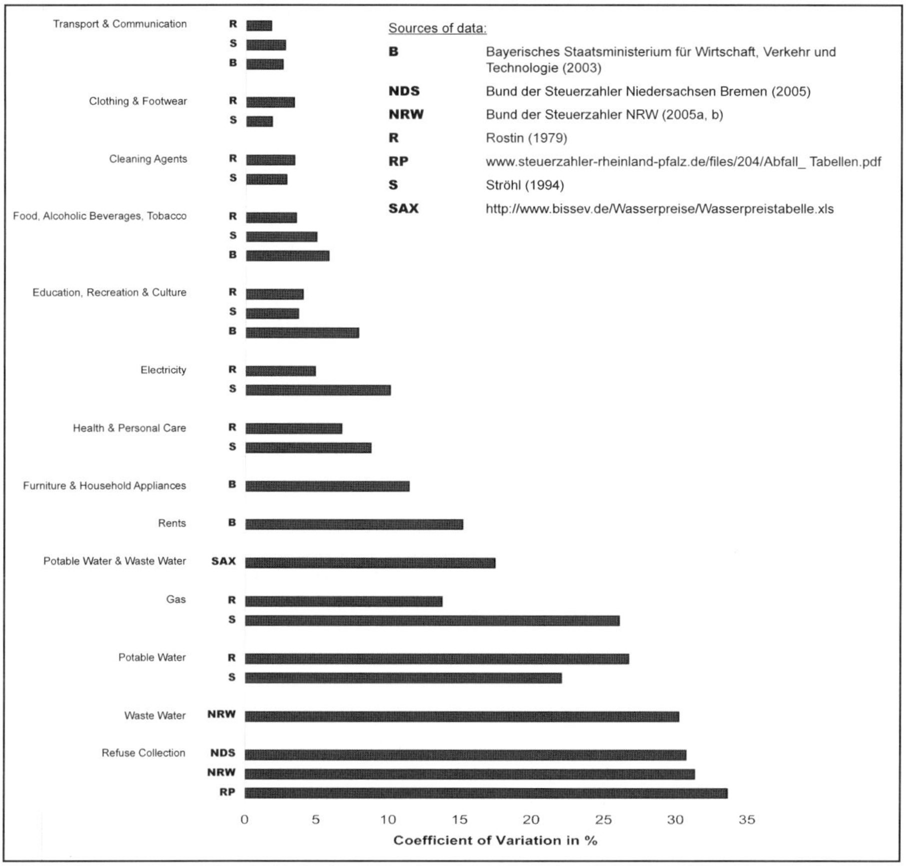

Fig.3: Extent of regional disparities in the prices of different groups of goods and services. Sources: Rostin (1979); Ströhl (1994); Bayerisches Staatsministerium für Wirtschaft, Verkehr und Technologie (2003); Bund der Steuerzahler Niedersachsen Bremen (2005); Bund der Steuerzahler NRW (2005a, b); http://www.bissev.de/Wasserpreise/Wasserpreistabelle.xls; www.steuerzahler-rheinland-pfalz.de/files/204/Abfall_Tabellen.pdf; author's own calculations.

of Western Australia[23], in 49 Japanese towns[24], in twelve regions in Great Britain[25] as well as in 38 American towns and regions[26] show a positive correlation between nominal and real income. Regression analyses based on all this data reveal that there is a significantly positive influence of nominal income on real income. This

[23] cf. http://www.dlgrd.wa.gov.au/Publications/Docs/indicatorsRegionDev_Economic.pdf and http://www.dlgrd.wa.gov.au/publications/docs/regionalpricesindex_section2.pdf

[24] Statistical Research and Training Institute (2007) and data from the Statistical Survey Department of Japan

[25] Wingfield/Fenwick/Smith (2005) and data from www.statistics.gov.uk/regionalsnapshot

[26] Aten (2006)

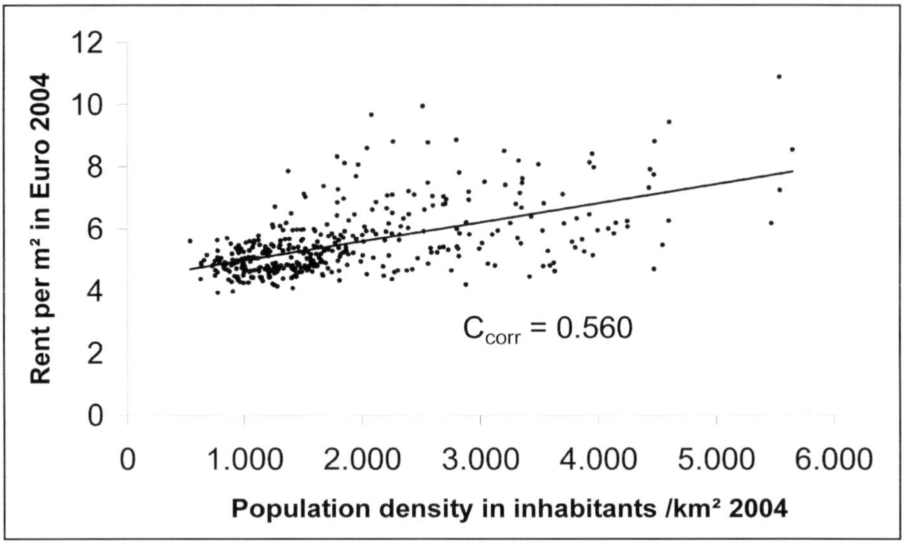

Fig.4: Influence of population density on rents. Source: BBR Housing Market Monitoring System, www.regionalstatistik.de, author's own calculations.

does not support the idea that COL compensate for the differences in nominal income completely – on the contrary, they remain even if real income is taken into consideration.

Second, the Bavarian data shows that the increase in COL between 1992 and 2002 is not uniform throughout the 21 communities observed. The coefficient of correlation between the values for the two years is neither high nor low (Ccorr = 0.629). Changes in the height of rents is the main reason for this unequal development, i.e. that some regions have become more expensive than others during the decade under consideration. The correlation between the indices without taking rents into account is Ccorr = 0.832. For the towns and cities which are included both in Werner Rostin (1979) and Gerd Ströhl (1994), the COL for the years 1978 and 1993 have a similar statistical relation (Ccorr = 0.784, again without rents).

Third, all three studies subdivide the COL index for different components in the basket of commodities. As a measure of the height of regional disparities, the coefficients of variation can be calculated (cf. Fig. 3, p 109).

It becomes apparent that some components in the basket of commodities have very low regional disparities in their prices, e.g., transport and communication, clothing and footwear, cleaning agents, food, alcoholic beverages and tobacco as well as education, recreation and culture.[27] These results are to be expected for

[27] The coefficient of variation in the Bavarian study is higher for this last component, because it does not include costs for education where regional disparities are relatively low.

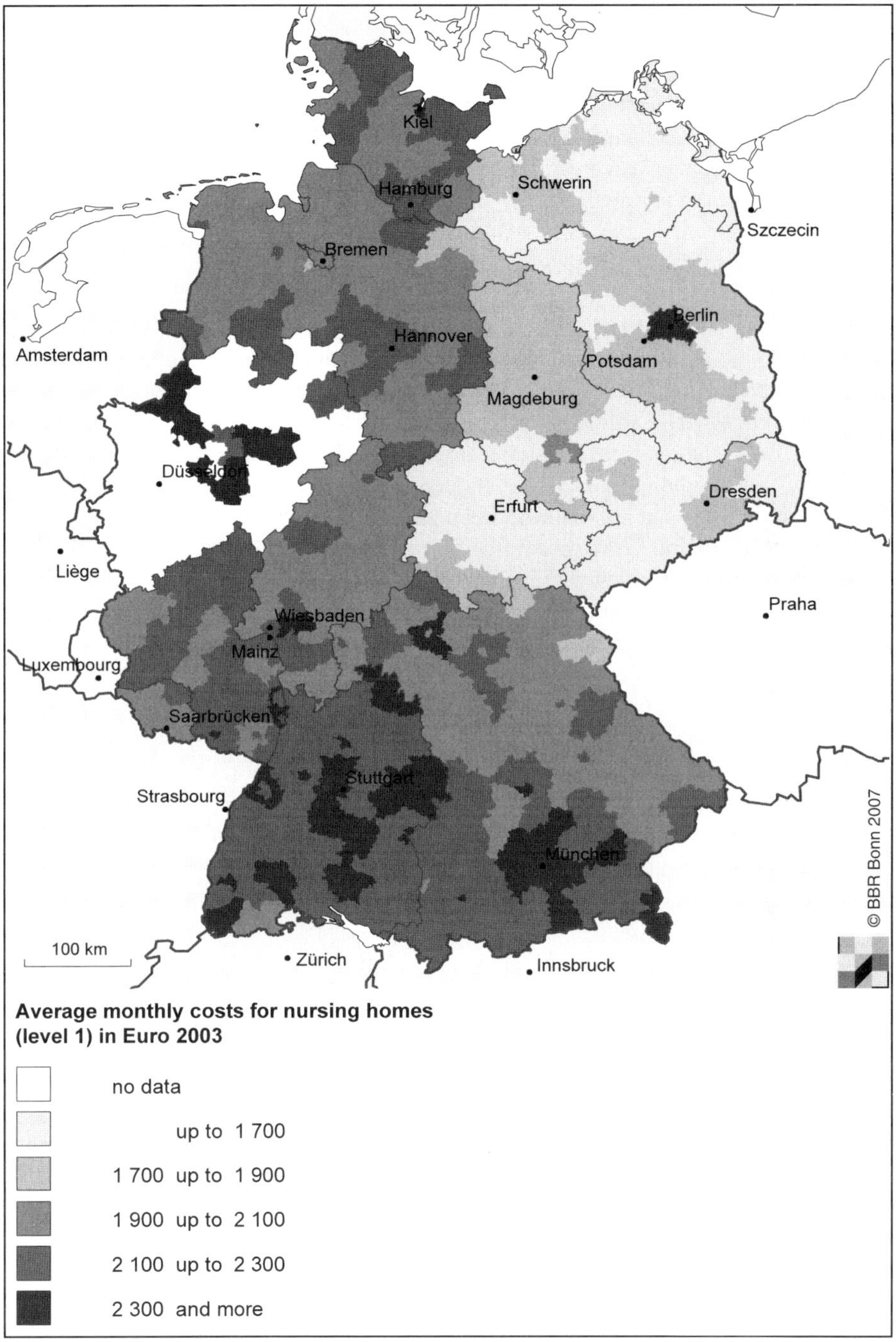

Fig.5: Costs for nursing homes. "Kreise" (counties) as of 31.12.2004. Source: BBK Pflegedatenbank and author´s own calculation.

transport and communication, as many prices are state-controlled. In the case of food and alcoholic beverages, the supermarket chains with their largely uniform pricing policies all over Germany, and the small margins in this line of business, lead to low regional variations in the costs of these items.[28] Refuse collection, water, gas and rents, however, show very high regional disparities, which is also not surprising because they are partially untradeable goods and occur where regional monopolies exist or existed at the time of the survey. This is especially the case for prices determined by local or regional administrative bodies or comparable institutions.

Fourth, the statement is very often made that COL rise with the number of inhabitants in a town, based on data both with and without rents. Not only do so all three more deeply analysed studies, but Andreas Mehnert and Michael Roos[29] also share this opinion. The Ministry of Internal Affairs and Communications provides a comparison of the regional consumer price index for Japanese towns according to the size of their population as well.[30] Indeed, there is a certain correlation between the two variables,[31] but this should be regarded more as a coincidence and less as a factor influencing COL, because it is not aggregated demand (if the number of persons living in a given region is interpreted in this way) but the relationship between aggregated demand and aggregated supply that determines the price.[32] Population density should be regarded as a more important impact instead of the absolute size of a population, which is reflected in higher correlation coefficients between COL and population density as well.[33] The reason is that available space, land prices and rents will become a major influence. Greater population densities imply, by definition, that there is less space per inhabitant, and thus space becomes a scarce and non-augmentable good and its price – whether for buying or renting – will rise (cf. Fig. 4, p 110). This has two effects on the consumer prices. On the one hand a direct impact, as the prices for housing rise. On the other hand, an indirect effect can be deduced: Suppliers of goods and services will also have to pay higher rents for office space, which will in turn lead to increased consumer prices.

[28] See also Ball/Fenwick (2003) who observed similar results in Great Britain.

[29] Mehnert (1997) p 21 and Roos (2003) p 10

[30] Ministry of Internal Affairs and Communications (2005) p 559

[31] e.g. Ccorr = 0.800 for the Bavarian data of 2002 with Munich, but only Ccorr = 0.405 without Munich

[32] Why should the COL rise when two neighbouring cities, e.g. Mannheim and Ludwigshafen, or two adjacent counties become one administrative unit?

[33] Ccorr = 0.822 for the Bavarian data of 2002 with Munich and Ccorr = 0.752 without Munich. The population density is not calculated based on the whole area within an administrative unit, but only on the built-up area ("Siedlungs- und Verkehrsfläche").

8.3 Some new Results

The Federal Office for Building and Regional Planning ("Bundesamt für Bauwesen und Raumordnung", BBR) does not only review the literature on this topic, it works on regional disparities of COL by collecting its own data as well. Internet-based surveys are the main source in this respect.[34] At the moment, the data is still too fragmentary to construct an index of COL. Therefore, the following paragraphs are devoted to four selected elements in the basket of commodities: nursing homes, local newspapers, English lessons in adult evening classes and heating oil.[35]

Nursing homes

Paying for nursing homes and nursery schools only adds up to 10.29 ‰ of a consumer's total expenditures. As an ample amount of information is available, however, nursing homes are used as an example of regional disparities in the COL. A CD-ROM of the BKK health insurance fund in Germany provided data on the monthly costs[36] – for a person in 7,323 nursing homes all over Germany except for some parts of North Rhine Westphalia – in 2004. It can be argued that the prices are mainly dominated by differences in quality. Indeed, it is to be expected that the least expensive one with a monthly cost of 750 Euro has less favourable facilities than the most expensive one with a monthly cost of 9,532 Euro. But an aggregation to the county level smoothes the results, as in the average 18 nursing homes exist within one county. The law of large numbers thus helps to remove major differences in quality – or at least reduces the impact they have on the average price in one county. This is, of course, not proof that they have completely vanished. The aggregation has the effect that the average costs per nursing home in these regions are reduced to 1,525 to 2,874 Euro. High-cost areas are the bigger cities like Munich, Stuttgart, Frankfurt (Main), Hamburg and Berlin. Furthermore, costs are high in large parts of southern Hesse, Baden-Wurttemberg and southern Bavaria. The northern regions of Germany have an average price, whereas the new "Länder" are comparatively inexpensive (cf. Fig. 5, p 111).

Local newspapers

Local and national daily newspapers are also elements in the basket of commodities. Together they make up an average of 8.8 ‰ of a consumer's total expenditures. An internet survey in 2006 provided the costs for 618 newspapers. In counties and towns with more than one local or regional newspaper the mean price was used. Prices for individual newspapers range from 12 Euro (per month) in the Saxonian county of Döbeln between Leipzig and Dresden to 28.75 Euro in Frankfurt (Main) and the

[34] I would like to thank Sabine Beißwenger for her excellent work in collecting the data.

[35] For information about the regional heights of rents see also: Bundesamt für Bauwesen und Raumordnung (2005) p 9 and for maps on taxi fees, fuel prices and household insurance tariffs: Kawka (2006a) pp 359ff.

[36] In "Pflegestufe 1"

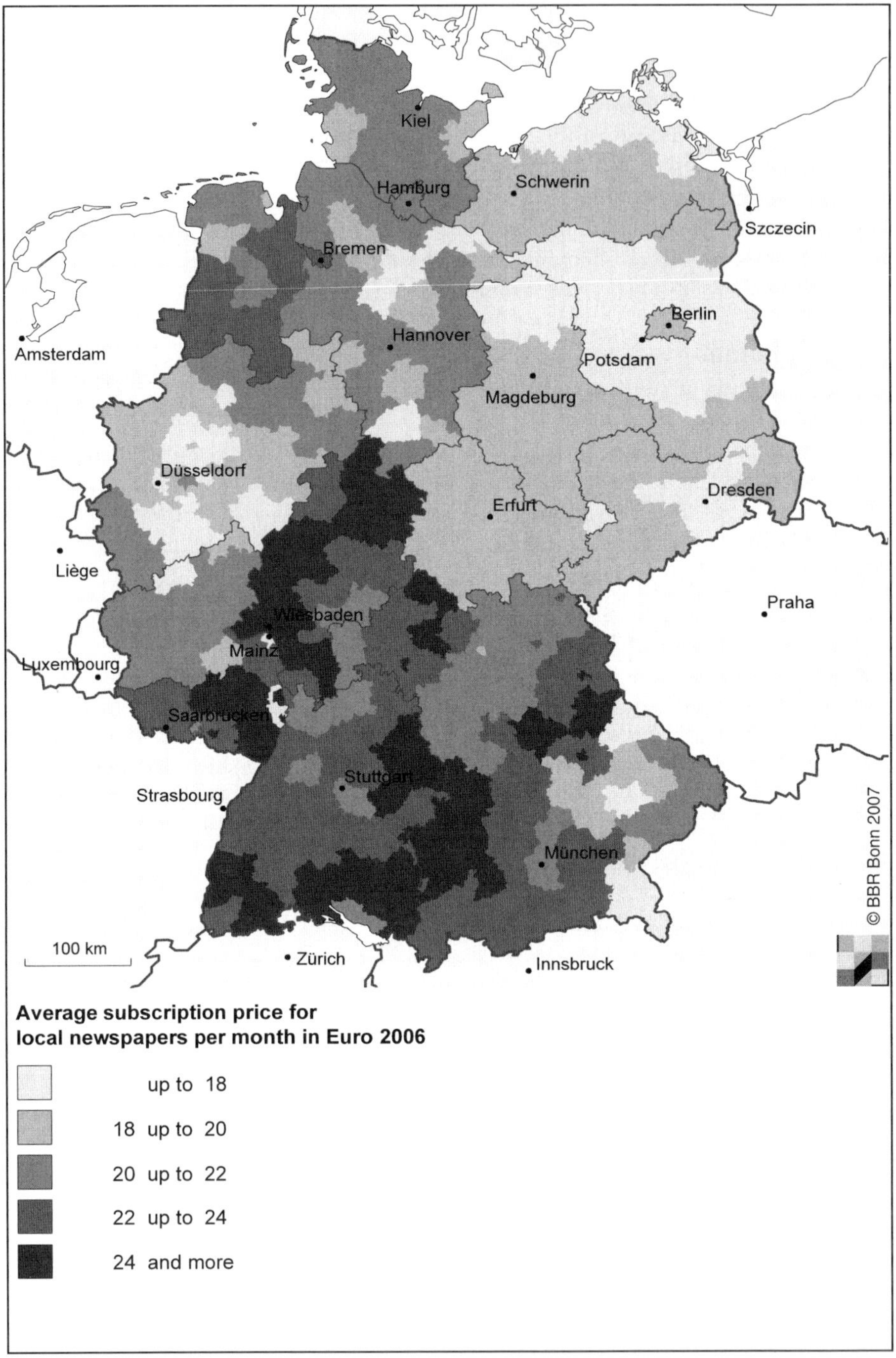

Fig.6: Prices of local or regional newspapers. "Kreise" (counties) as of 31.12.2004. Source: Internet survey, author´s own calculation.

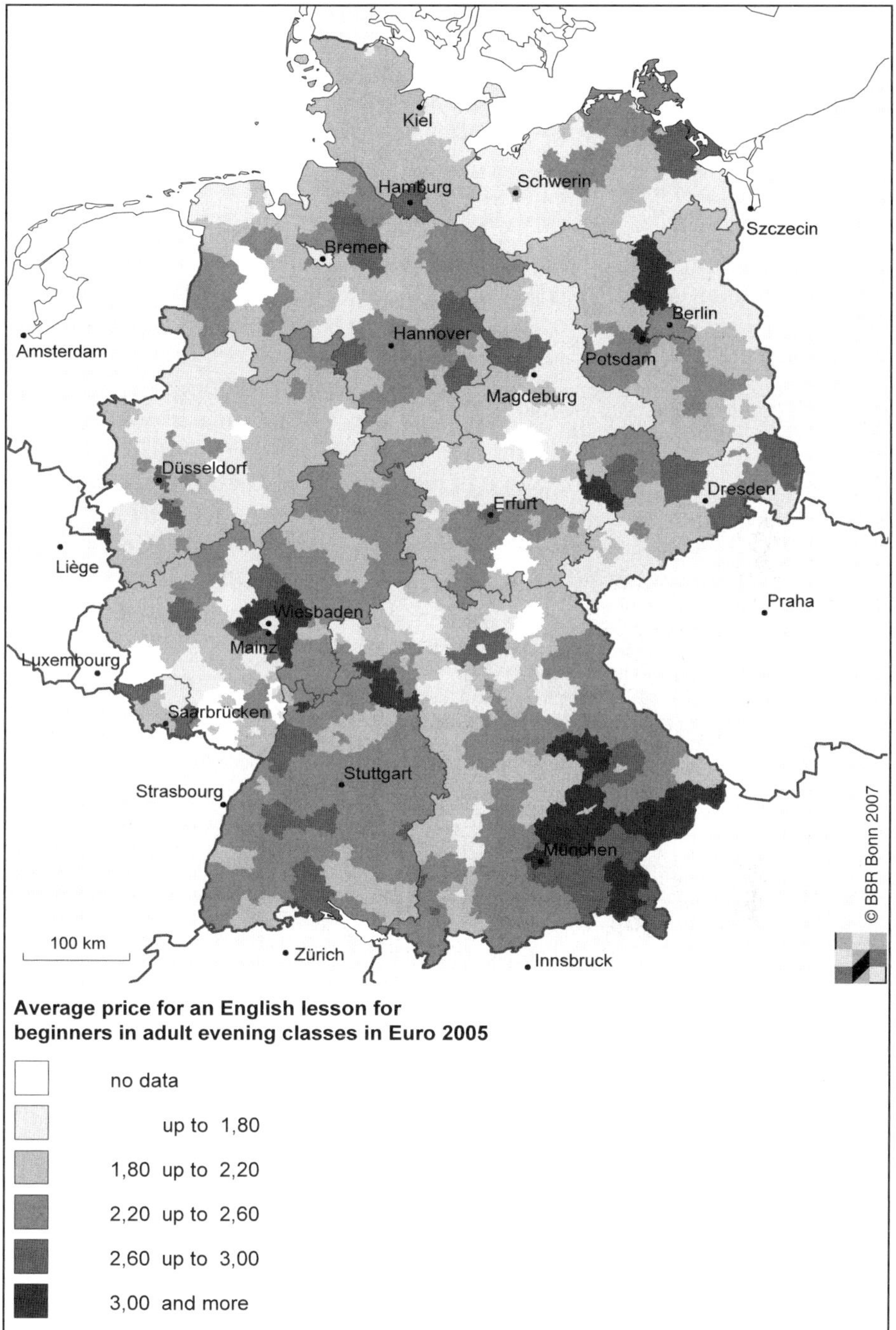

Fig.7: Prices for English lessons in adult evening classes. "Kreise" (counties) as of 31.12.2004. Source: Internet survey, author´s own calculation.

surrounding area. As one newspaper may serve more than one county, sometimes with regionally adapted content but often with the same price, it can be expected that the price of a local newspaper does not exactly mirror the size of disposable income on the regional level. Nevertheless, in those states with comparatively high incomes, i.e. Hesse, Baden-Wurttemberg and Bavaria, the prices are the highest as well (cf. Fig. 6, p 114).

English lessons in adult evening classes

Tertiary education, at 1.72 ‰ makes up only a very small share of the expenditures of a representative consumer. An English course for beginners is taken as an example for regional price disparities in this respect, because this type of course is offered in nearly every adult evening class ("Volkshochschule") in Germany. An internet survey generated the prices of 935 courses. Where no English class was available, a French course was taken as a substitute. Bernburg in Saxony-Anhalt is the least expensive region with a cost of only 1.14 Euro per lesson. The most expensive class is found in Oberhavel in Brandenburg, and costs 6.29 Euro per lesson. Interregional disparities are great, the coefficient of variation (CV) is 21.8%. In comparison: the CV for newspapers is merely 12.6 ‰ and for nursing homes 11.5%. Fig. 7 (at page 115) shows no clear spatial pattern, but once again there is a tendency for higher prices in the states of Hesse, Baden-Wurttemberg and Bavaria. Courses are very expensive in south-eastern Bavaria and in Frankfort (Main) and surroundings.

Heating oil

The average consumer pays 7.9 ‰ of his total expenditures for heating oil. The internet page www.fastenergy.de provides excellent information about prices in German counties. Here, the costs for 100 litres of heating oil on 21 June 2006 are taken as a reference, when a volume of 3,000 litres is bought. The data used here is based upon a survey of 719 offers for oil throughout Germany. The lowest price is in Heinsberg, in North Rhine-Westfalia, at 58.44 Euro, whereas the same amount costs 66.87 Euro in Waldshut in Baden-Wurttemberg, or 14% more. This does not mean that the differences persist over time. The spatial pattern of fuel prices can change quickly, and an analysis based on only one day of the year should be regarded only as a snapshot. The CV is extremely low at only 2.2%. Fig. 8 (on page 118) shows a rather coincidental spatial pattern of the costs. At the date in question, southern Germany seemed to be more expensive than the northern part, although, for example, heating oil is quite expensive in eastern Mecklenburg-Western Pomerania, too. There are also some comparatively inexpensive 'islands' in the south.

8.4 Conclusions

As mentioned in the introduction, this article cannot definitively answer questions of how COL affect the spatial pattern of real income. Three scenarios are possible: **(1)** COL overcompensate so that regions with a low nominal income have a high real income, **(2)** COL balance the regional disparities in nominal income, and **(3)** the ranking order of the regions according to real income is the same as with nominal income.

The survey of literature gives no hints concerning the first scenario. It is thus very doubtful that the following quotation is appropriate in this context: "But many that are first shall be last; and the last shall be first" (Matthew 19, 30), to once again quote the Bible, as was done in last year's German Annual of Spatial Research and Policy.[37] For spatial planning, which still has the aim of achieving equivalent living conditions, this result would be undesirable too due to the creation of new regional disparities.

It can be taken for granted that regional disparities in real income are lower than in nominal income, as COL balance them to a certain extent. It is still an open question whether the second or the third scenario is more plausible. Findings based on the Bavarian data somehow support the second scenario – although statistically not very strongly, and are dependent on the inclusion of Munich in the sample. The second scenario, the most desirable result for regional planning, would imply that the goal of equivalent living conditions has already been achieved with regard to average disposable income – thanks to market forces. International findings quoted back the third hypothesis. This could be regarded as a preferable outcome too, as long as the differences in real income are not too high, which is quite likely. The remaining disparities would serve as an incentive to improve the material standard of living for inhabitants in less wealthy regions and open options for regional policy to try to converge them with more successful regions. In any case, it is a certain relief for regional policy that market forces promote their efforts in achieving equivalent living conditions.

[37] Kawka (2006b) p 51

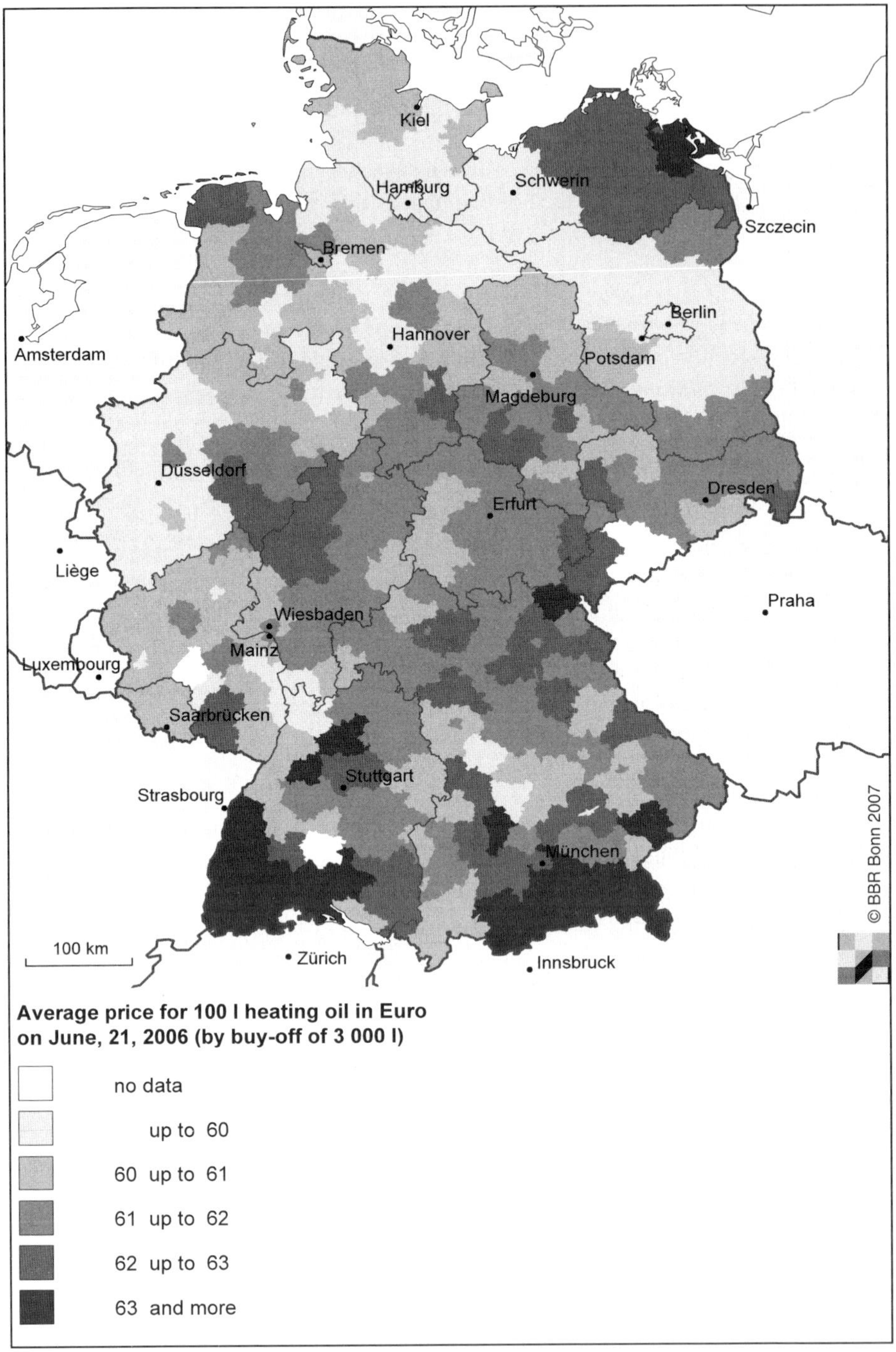

Fig.8: Prices for heating oil. "Kreise" (counties) as of 31.12.2004. Source: www.fastenergy.de, author´s own calculation.

References

Angermann O (1989) Vergleich des Verbraucherpreisniveaus in Bonn, Karlsruhe, München und Berlin (West). In: *Wirtschaft und Statistik*, Issue 04/1989, pp 258-261

Aten B (2006) Interarea price levels: an experimental methodology. In: *Monthly Labor Review*, 09/2006, pp 47-61

Ball A, Fenwick D (ca. 2003) Relative Regional Consumer Price Levels in 2003. Available online: www.statistics.gov.uk/articles/nojournal/Regional_Price_Levels_2003_revised.pdf, visited on 3.1.2005

Bayrisches Staatsministerium für Wirtschaft, Verkehr und Technologie (2003) Die reale Kaufkraft in Bayern 2002. München

Bundesamt für Bauswesen und Raumordnung (BBR) (2005) Raumordnungsbericht 2005. Berichte Vol. 21, Bonn

Bund der Steuerzahler Niedersachsen Bremen (2005) Niedersächsische Abfallgebühren im Vergleich. Available online: http://www.steuerzahler-niedersachsen-bremen.de/file.php?f=178/Abfallgebuehrenstudie_geaenderte_Fassung_04-08-05.pdf, visited on 26.7.2006

Bund der Steuerzahler NRW (2005) Gebührensituation in NRW 2005 Anlage 1: Abfallgebühren. Available online: http://www.steuerzahler-nrw.de/download/abfall2005neu7.pdf, visited on 26.7.2006

Bund der Steuerzahler NRW (2005a) Gebührensituation in NRW 2005 Anlage 2: Abwassergebühren. Available online: http://www.steuerzahler-nrw.de/download/abwasser2005.pdf, visited on 26.7.2006

Buschle S (1997) Interregionale Preisunterschiede im Lebensmitteleinzelhandel. Materialien des Zentrums für regionale Entwicklungsforschung der Justus-Liebig-Universität Giessen, Vol. 42, Giessen

Chowhan J, Prud'Homme M (2004) City comparisons of shelter costs in Canada: A hedonic approach. Statistics Canada Research Paper. Ottawa

Cullison WE (1984) Equalizing Regional Differences in Wages: A Study of Wages and Migration in the South and Other Regions. In: *Economic Review*, Issue: May/June 1984, pp 20-33

Curran H, Wolman E, Hill S, Furdell K (2005) Economic Wellbeing and Where We Live: Accounting for Geographic Cost-of-living Differences. Draft Version, without place of publication

Czajka S, Kott K (2006) Konsumausgaben privater Haushalte für Nahrungsmittel, Getränke und Tabakwaren 2003. In: *Wirtschaft und Statistik*, Issue 06/2006, pp 630-643

Engels E (2003) Zur Anwendung hedonischer Methoden auch bei dezentralen Positionen der Verbraucherpreisstatistik. In: *Berliner Statistik*, Issue 11/03, pp 446-449

Evers H, Schmuckler H, Teich G (1963) Regionale Unterschiede in den Lebenshaltungskosten. In: *Mitteilungen aus dem Institut für Raumforschung*, Issue 48, Bad Godesberg

Fenwick D, O'Donoghue J (2003) Developing Estimates of Relative Regional Consumer Price Levels. Available online: www.statistics.gov.uk/articles/nojournal/Regional_Consumer_Price_levels.pdf, visited on 04.01.2005

Gesellschaft für Konsumgüterforschung (1982) Die reale Kaufkraft in Bayern. Nürnberg

Jolliffe D (2006) Poverty, Prices, and Place: How Sensitive is the Spatial Distribution of Poverty to Cost of Living Adjustments? In: Economic Inquiry, Vol. 44, No. 2, pp 296-310

Kawka R (2006a) Unterschiede in den Lebenshaltungskosten: Ein Ausgleich für regionale Disparitäten? In: *Informationen zur Raumentwicklung*, Issue 06/07 2006, pp 355-361

Kawka R (2006b) Regional Disparities in the GDR – Do They Still Matter? In: Lentz S (ed.) German Annual of Spatial Research and Policy Restructuring Eastern Germany, Heidelberg-Berlin, pp. 39-53

Kokoski M, Cardiff P, Moulton B (1994) Interarea Price Indices for Consumer Goods and Services: A Hedonic Approach Using CPI Data. U.S. Department of Labour BLS, Working Paper 256, Washington D.C.

Kreisverwaltung Ahrweiler (1999) Besser leben Kreis Ahrweiler. Bad Neuenahr-Ahrweiler

Mehnert A (1997) Das reale Einkommen im interregionalen Vergleich – Der Einfluss des Preisniveaus auf die regionalen Einkommensdisparitäten unter Anwendung des statistischen und des ökonomischen Indexkonzeptes. IWH-Diskussionspapiere No. 66, Halle (Saale)

Ministry of Internal Affairs and Communications (2005) Japan Statistical Yearbook 2006, Tokyo

Office of Economic and Statistical Research Queensland Treasury (2001) Index of Retail Prices in Regional Centres October 2001, Brisbane

Persson J (1994) Convergence in Per Capita Income and Migration across the Swedish Counties 1906-1990. Seminar Papers, Institute for International Economic Studies, Stockholm University No. 601, Stockholm

Rostin W (1979) Zwischenörtlicher Vergleich des Verbraucherpreisniveaus in 31 Städten. In: *Wirtschaft und Statistik*, Issue 06/79, pp 403-410

Roos M (2003) Regional price levels in Germany. Dortmund. Available online: www.ersa.org/ersaconfs/ersa03/cdrom/papers/511.pdf, visited on 4.1.2005

Siminski P, Saunders P (2003) Low-income before and after Housing Costs, Comparing Australia's Regions. Available online: www.sprc.unsw.edu.au/ASPC2003/papers/Paper222.pdf, visited on 10.1.2005

Statistical Research and Training Institute (2007) Japan Statistical Yearbook 2007. Tokyo

Statistisches Bundesamt (2004a) Einnahmen und Ausgaben privater Haushalte. Presseexemplar. Wiesbaden

Statistisches Bundesamt (2004b) Regionale Unterschiede in den Lebenshaltungskosten. In: *Wirtschaft und Statistik*, Issue 09/2004, p 1050

Ströhl G (1994) Zwischenörtlicher Vergleich des Verbraucherpreisniveaus in 50 Städten. In: *Wirtschaft und Statistik*, Issue 06/1994, pp 415-434

Struff R, Wilamowitz-Moellendorf v. U (1983) Räumliche Einkommens-, Preis- und Verbrauchsunterschiede in der Bundesrepublik Deutschland. *Schriftenreihe der Forschungsgesellschaft für Agrarpolitik und Agrarsoziologie*, Vol. 262, Bonn

Wingfield D, Fenwick D, Smith K (2005) Relative regional consumer price levels in 2004. In: *Economic Trends*, Issue 615, 02/2005, pp 36-45

Online Sources:

www.bissev.de/Wasserpreise/Wasserpreistabelle.xls (visited on 14.3.2005) Wasserpreise in Sachsen

www.bverfg.de/entscheidungen/rs20070306_2bvr055604.html (visited on 7.3.2007) Leitsatz zum Urteil des Zweiten Senats vom 6. März 2007 - 2 BvR 556/04 - Zur Frage, ob das Alimentationsprinzip den Besoldungsgesetzgeber verpflichtet, regional unterschiedliche Lebenshaltungskosten auszugleichen

www.coli.org (visited on 21.7.2006) ACCRA Cost of Living Index.

www.dlgrd.wa.gov.au/Publications/Docs/indicatorsRegionDev_Economic.pdf (visited on 19.1.2007) Regional performance – economic

www.dlgrd.wa.gov.au/publications/docs/regionalpricesindex_section2.pdf (visited on 27.2.2007) State and Commodity Analysis

www.fastenergy.de (visited on 28.7.2006) Heizöl günstig bestellen, Heizölpreise online

www.regionalstatistik.de (visited on 14.7.2006) GENESIS-Online regional – Das statistische Informationssystem des Bundes und der Länder

www.statistics.gov.uk/regionalsnapshot (visited on 28.2.2007) Regional Snapshot.

www.steuerzahler-nrw.de/download/abfall2005neu7.pdf (visited on 17.7.2006) Gebührensituation in NRW 2005, Anlage 1: Abfallgebühren

www.steuerzahler-nrw.de/download/abwasser2005.pdf (visited on 17.7.2006) Gebührensituation in NRW 2005, Anlage 2: Abwassergebühren

www.steuerzahler-rheinland-pfalz.de/files/204/Abfall_Tabellen.pdf (visited on 3.8.2006) Abfallgebühren in Rheinland-Pfalz. Vergleich der Abfallgebühren 2003 zwischen Normal- und Sparhaushalt – Landkreise

9 Territorial Cohesion and Transboundary Governance: Insights from the Polish-German and the Czech-German Borders

Markus Leibenath, Robert Knippschild

Introduction

Territorial cohesion as a challenge for transboundary governance

The term territorial cohesion has evolved as a new buzz-word among policy-makers and academics dealing with European spatial development. After it was officially introduced in the Amsterdam Treaty establishing the European Community[1], the notion soon gained increasing attention in policy discourse. For instance, informal ministerial meetings on territorial cohesion were organised at the EU level in 2004 and 2005. In 2005 a special issue of the British scientific journal Town Planning Review as well as a track of the annual congress of the Association of European Schools of Planning were also dedicated to this theme.

A neutral observer, however, could sometimes get the impression that many scholars are struggling hard to grasp the meaning of territorial cohesion and to transfer it from "a vague notion into a clear idea."[2] Some analysts even question whether territorial cohesion represents an innovative approach or rather "old (French) wine in new bottles",[3] and warn not to overemphasize instrumental interpretations.[4]

So what is territorial cohesion all about? Basically, two meanings can be distinguished. First, territorial cohesion implies the harmonious and balanced development of the EU as a whole. This reflects the old concern about territorial inequities and the maintenance of services in rural and peripheral areas. Or in one sentence, "People should not be disadvantaged by wherever they happen to live or work in the Union."[5] This objective is in line with the French-style interventionist regional economic approach, but also, for example, with the claim of ensuring equal living conditions in all parts of the country, which can be found in Articles 72-2 and 91a-1 of the German Constitutional Law.[6]

---

[1] OJEC (1997/C 340/35) – article 16; see also EU-COM (2001) p 29

[2] Gustedt (2005) p 1

[3] Faludi (2003) p 1

[4] Gualini (2005) p 1

[5] EU-COM (2004) p 27

[6] BGBl (1949), III 100-1; EU-COM (2003) p 40; Faludi (2003) p 3

The second meaning of territorial cohesion centres on co-ordination "between regional policy and sector policies with a substantial territorial impact"[7] and of "national policies particularly across borders."[8] It represents a long-term approach "at, and across, different policy levels (regional, national, transnational and EU) and across sectors (agriculture, transport, environment, regional-economic development, competition, etc.)."[9] The European Commission[10] also wants to encourage co-operation between regions within the framework of territorial cohesion.

The rationale behind the term territorial cohesion is twofold. From a discourse analysis point of view territorial cohesion is related to the Lisbon strategy and its underlying growth and jobs agenda. Thus, the emphasis on territorial cohesion represents a shift of emphasis away from a comprehensive understanding of sustainable development and towards a priority on competitiveness and economic performance. Another aspect refers to the competence of the European Commission. Cohesion policy is part of the EU's regional policy and therefore at the heart of its sphere of influence. By contrast, spatial development at the European level does not fall into the range of formal EU competencies and cannot be controlled by the Commission. This is perhaps one of the major reasons why the Commission did not support an ESDP follow-up whatsoever but concentrated rather on the territorial cohesion approach.[11]

To sum up, territorial cohesion as a political objective focuses on the reduction of imbalances between different parts of EU territory. To this end, European, national, and regional policies which have a substantial territorial impact shall be co-ordinated both between different sectors and political-administrative levels of policy-making, as well as across national borders in order to avoid or reduce adverse territorial effects. With regard to the Lisbon strategy, the imperative of territorial cohesion also includes strengthening the EU's competitiveness at global scale, enhancing the territorial capital and potential of all EU regions, and promoting territorial integration.[12]

Co-ordination of spatially relevant policies across national borders is a typical issue of transboundary governance, understood as addressing spatial concerns that transcend traditional national problem-solving capacities in a politically meaningful way.[13]

The intention of this article is to analyse the mechanisms of transboundary co-ordination of spatial policies at the Polish-German and Czech-German borders. In doing so, we want to answer the following research questions: To what extent do current approaches towards the transboundary co-ordination of spatial policies

[7] EU-COM (2003) p 40
[8] DG Regio (2004) p 9
[9] Luxembourg presidency of the Council of the European Union (2005) p 4
[10] EU-COM (2004) p 27
[11] Faludi (2003) p 3
[12] Luxembourg presidency of the Council of the European Union (2005) p I
[13] Scott (2002) p 136

already contribute to territorial cohesion? What are crucial influencing factors for the success of transboundary co-ordination, especially with regard to the integration of different levels of policy-making and to different types of governance architecture? And finally: What conclusions can be drawn with regard to future transboundary co-operation? We start with some theoretical considerations about transboundary territorial governance. We then present three case studies concerning the transboundary co-ordination of spatial policies at communal, regional, and intergovernmental levels, based on a brief description of the spatial situation and the transboundary institutional framework within the respective regions. Finally, we compare the three cases and draw conclusions with regard to the theoretical discussion presented in the following chapter.

The particular cases have been chosen for analysis not only because they represent the different levels of transboundary governance – national, regional and local – and allow for a comparison to be made. They also highlight the specific context for transboundary governance at a former external border of the EU because Poland and the Czech Republic did not join the Union until May 2004. Moreover, we had the opportunity to be actively involved as researchers and consultants in the three respective projects, which gave us the opportunity to closely monitor the processes and to get access to many unpublished documents.

9.1 Transboundary Governance Concerning Policies with a Spatial Impact

Coping with asymmetries

It is an overall objective of European integration policies to overcome borders and to create a coherent territory without internal frontiers according to Articles 14 and 154 of the EC Treaty[14] and Article I-3 of the European Constitution.[15] The EU has also passed several pieces of legislation in recent years that include provisions for transboundary co-ordination of spatial policies and procedures, for example the Water Framework Directive[16] or the directive on environmental assessments of plans and programmes.[17] Furthermore, there are efforts towards the transboundary co-ordination of spatial policies at the European level with regard to trans-European networks and flood-risk management. However, in many other fields of spatially relevant policies like housing, nature conservation, and regional planning no supranational or European institutional framework is available. This is why actors who strive to coordinate such policies have to cope with the asymmetry of their respective national institutions and to create cross-border institutions on a case-by-case basis.

[14] OJEC (2002/C 325/33)
[15] OJEC (2004/C 310/1)
[16] OJEC (2000/L 327/1)
[17] OJEC (2001/L 197/30)

The term 'institutional asymmetries' sounds abstract. What does it mean with regard to the transboundary co-ordination of spatial policies? In the social sciences, institutions are understood as "the ensemble of norms, rules and practices which structure action in social contexts."[18] Typical institutions in the field of spatial planning are legal frameworks that define the types and contents of planning documents, establish planning procedures including, for instance, requirements for public participation, and specify the role and competencies of public authorities at different levels. These institutions are closely related to the overall administrative structure of a state. In the case of a border between countries with very similar overall political-administrative structures, the institutional asymmetries will be less articulated and it will be easier to achieve common understanding and effective policy co-ordination. However, even in these cases there may be asymmetries which can be subsumed under the heading 'culture'. This refers not only to the easily visible differences of languages, but also to planning traditions and cultures as well as to more generic cultural differences between nations. The latter concerns issues like time management, the relative importance of oral as opposed to written agreements, and the role of informal, private contacts between co-operation partners, to mention just a few.[19]

There may also be other things beside institutional asymmetries between two countries which are relevant to the co-ordination of spatial policies. These are the cognitive-informational and the economic-technological framework conditions.[20] The first describes the conditions under which knowledge is generated, distributed, interpreted, and applied. Spatial planners in one country, for example, might dispose of sophisticated data bases and spatial monitoring schemes whereas their counterparts from a neighbouring country do not. Economic-technological framework conditions include the overall wealth of a nation as well as the distribution of income and the situation of public budgets. In sum, both institutional and other structural framework conditions determine the way spatial problems are perceived and constituted, but also the actors' capacities for action and their orientation.[21] As framework conditions usually differ from one state to another, various asymmetries arise which hamper the transboundary co-ordination of spatial policies.

Contents and structures of the the transboundary co-ordination of spatial policies

As the above discussion has showed, the transboundary co-ordination of spatial policies is a demanding business. It is not a new one, though. Typically, four different types or levels of transboundary co-ordination can be distinguished, all of which do not necessarily follow any inherent teleology:

[18] Healey (2006) p 302

[19] For more details, see e.g. Schroll-Machl/Nový (2003)

[20] Jänicke (2002) pp 5f

[21] Scharpf (1997) pp 58-66

- The transboundary exchange of information on national legal frameworks, on national procedures and competencies, on perceptions of transboundary spatial issues, as well as on ongoing planning activities. Such an exchange of information can take the form of a one-time activity or can be practiced on a more continuous and regular basis.
- The transboundary co-ordination of individual projects or plans with potentially significant impacts on the neighbouring state. Examples would be the construction of power plants or large commercial or industrial sites close to the border.
- Co-ordinated agenda-setting and the joint preparation of spatial plans, strategies and visions.
- The co-ordinated representation of common interests in opposition to other agencies or organisations, for example the EU or the European Commission.

The first level can be regarded as a precondition for the activities of the other two types: Without knowledge about structures and without direct contacts between individuals and organisations, policy co-ordination cannot be effected. These first-level activities usually lead to the rapprochement of perceptions and mental frameworks and to a better mutual understanding of planning cultures. This is what Zonneveld[22] refers to as the "broad interpretation of transnationality." By contrast, from a narrow view on transnationality, actors have to concentrate on issues where planning in the sense of programming – i.e., the actual physical implementation of spatial plans or agreements – is intended. This can apply to transport infrastructure, ecological networks or issues of water management.[23] Many efforts regarding the transboundary co-ordination of spatial developments, however, start with the ambition of complying with the narrow interpretation of transnationality and end up with results that do not go beyond "planning as communication."[24]

The broad interpretation of transnationality corresponds to the notion of planning as communication, with the function of providing interpretative frameworks and of changing the reference frameworks of those actors involved. Transboundary co-operation in the field of spatial policies that is in line with the broad type of transnationality contributes to institutional capacity-building. Generally speaking, institutional capacity refers to the overall quality of the collection of relational networks in a place. It allows rapid mobilization to deal with new circumstances and enables flexible responses to be designed and developed.[25] Thus, transnational planning as communication is the prerequisite for more advanced forms of the cross-border co-ordination of spatial policies. Social capital, and especially trust, also plays a crucial role for transboundary institutional capacity.[26]

[22] Zonneveld (2005) p 148
[23] Zonneveld (2005) p 148 and p 151
[24] de Vries (2004) p 10
[25] Healey (1997) p 61; Healey (1998) p 1541
[26] Klein-Hitpaß et al. (2006)

What are the typical structures of the transboundary co-ordination of spatial policies? Three basic types of organizational arrangements can be distinguished:

- Project-based co-operation, for instance within the framework of INTERREG,
- permanent transboundary structures like city networks, Euroregions, or intergovernmental commissions which often establish more specialized thematic working groups, and
- transboundary co-operation as part of the regular day-to-day planning and licensing routines of bureaucrats and stakeholders.

It seems as if permanent organisational structures were a precondition to the evolution of day-to-day co-ordination routines. The intergovernmental commissions for spatial planning between Germany and its western and southern neighbour states were initiated in the early 1970s. One of their core activities was to adopt recommendations on how to precisely organise the transboundary co-ordination of those plans and projects that could considerably affect the other state's territory. Not surprisingly, the activity of most of these commissions slowed down once such agreements had been reached.

Functional or territorial organisation?

Blatter[27] has coined the antagonism of "spaces of place" and "spaces of flows" with regard to cross-border co-operation. "Spaces of place" refers to a territory-centred – or one could also say a jurisdiction-centred – architecture of governance which is dominated by members of the public sector and which bundles many tasks within one or a few jurisdictions. The respective governance structures tend to be rather formalized and quite stable with respect to time and space. The prototype of such an organisation is a traditional intergovernmental commission which aims at centralizing cross-border information flows. "Space of flows" marks the opposite end of the scale, where function-centred architectures of governance can be found. In this case, direct contacts between sub-national actors of various types as well as the inclusion of non-public-sector organizations prevail.[28]

These terminological distinctions are congruent with the observation that borders between sovereign states are not any more the one single "Westphalian"[29] line[30] which separates two countries in all respects. Instead, there is an overlap of different functional spaces which are each delimited individually. As a consequence, economic, social, legal, political, and identity spaces are to an ever greater extent

[27] Blatter (2004)

[28] Blatter (2004) p 533; Castells (1996) p 29; Fürst (2003) p 442

[29] The Peace of Westphalia from 1648 officially ended the Thirty Years' War and paved the way to a mosaic of sovereign, completely autonomous micro-states all across Germany.

[30] Brunet-Jailly (2004) p 1

bounded separately.[31] This is due to the growing complexity of modern societies and their functional segmentation, which makes it impossible to control all external relationships with one territorial borderline.

It is largely unclear, however, what type of governance architecture is needed for the transboundary co-ordination of spatial policies. Certainly the issue is less pressing if 'only' transnationality in the broad sense is desired, because mutual policy learning and reframing processes do not require functional interdependence. But what about the narrow interpretation of transnationality? Does transboundary policy co-ordination, which is meant to accomplish solutions for transboundary spatial decision-making problems of a physical nature, require a functional or rather a territorial architecture? Or does it not make any difference at all?

These questions form one of the parameters for the following case studies from the Polish-German and Czech-German borders. The case studies are also supposed to shed light on the influence of transboundary asymmetries and on ways of dealing with them, as well as on the interplay of the contents and structures of the transboundary co-ordination of spatial policies.

9.2 Empirical Lenses on the Polish-German and Czech-German Borders

In the centre of Europe but yet on the periphery

When discussing territorial cohesion and transboundary governance in borderlands, one has to take into account the structural situation of the regions concerned. This refers to their location and accessibility, to socio-economic parameters, transboundary interdependencies and the need for policy co-ordination as well as to the institutional framework conditions.

Most parts of the border regions along the Polish-German and Czech-German borders are located far from metropolitan agglomerations. The transboundary transport infrastructure also has severe gaps that further worsen accessibility in comparison to other peripheral regions. In particular public transport is hardly attractive for transboundary journeys. Overall travel times when using public transport are extremely long. A journey of not more than 120 kilometres between regional centres in the Euroregion Neiße-Nysa-Nisa Jelenia Góra, Bautzen and Liberec can take several hours. The frequency of service is not particularly satisfying as well, with only one or two connections per day. In some areas, however, transboundary transport connections are fairly good, for instance between Prague and Dresden.

There are considerable structural differences within the German-Polish-Czech borderland. The density of population varies between 73 inhabitants per square kilometre in the Polish Lubuskie Voivodeship ("Województwo Lubuskie") to 264 inhabitants per square kilometre in the district of Chemnitz ("Regierungsbezirk

[31] Christiansen/Jørgensen (2000) p 62

Chemnitz"). The ageing structure and the level of education also differ. While the proportion of population over 65 years in the districts of Ústí nad Labem ("Ústecký kraj") and Karlovy Vary ("Karlovarský kraj") is 12.9%, it amounts to 19,1% the district of Chemnitz ("Regierungsbezirk Chemnitz") in the year 2000. Almost 31% of the population between 25 and 64 years in the administrative district of Dresden ("Regierungsbezirk Dresden") had an academic degree while it was only 7.2% in the districts of Ústí nad Labem ("Ústecký kraj") and Karlovy Vary ("Karlovarský kraj") in 2002. The unemployment rate varies in the year 2002 between 5.4% in the district of Liberec (Liberecký kraj) and 26.1% in the Lower SilesiaVoivodeship ("Województwo Dolnośląskie") or 20.8% in the district of Dresden ("Regierungsbezirk Dresden"). The comparison of per capita income again shows the harsh differences between the eastern and the western sides of the borders. While the Lubuskie Voivodeship ("Województwo Lubuskie") had a per capita GDP of 60.1% below the EU 25 average in 2001, the Bavarian district of Oberpfalz ("Regierungsbezirk Oberpfalz") was 9.8% above the EU 25 average. These differences, however tend to shrink in a long-term perspective. While there was a growth of GDP in the districts of Dresden and Chemnitz ("Regierungsbezirk Dresden" and "Regierungsbezirk Chemnitz") of only 1.3% between 1995 and 2001, it was 6.3% during the same period in the Lower Silesia Voivodship ("Województwo Dolnośląskie").[32]

At the same time, the need for transboundary policy co-ordination in these border regions is obvious. River basins, mountain ranges, habitats, transport corridors and infrastructure facilities stretch across the national borders. Integrated planning or development across borders, for instance in the Ore Mountains along the German-Polish border, in the Oder and Neiße valley along the German-Polish border, or within the Paneuropean Transport Corridors II, III and IV, require transboundary co-ordination. With regard to institutional framework conditions, several bilateral treaties should be mentioned. The most important ones are:

- The 1991 German-Polish treaty on neighbourliness and amicable co-operation[33] and
- the 1992 German-Czech-Slovak treaty on neighbourliness and friendly co-operation. [34]

Although a lot more bilateral treaties exist, for example concerning co-operation in environmental protection and water management, there is no legal basis for the transfer of formal competencies from national authorities to transboundary organisations. By contrast, such arrangements are possible at Germany's western borders thanks to the Agreement of Karlsruhe.[35]

[32] EU-COM (2004) pp 188-200
[33] BGBl (1991) II p 1315
[34] BGBl (1992) II p 462
[35] BGBl (1997) II p 1159

At the national level, two intergovernmental committees are concerned with spatial planning issues. The first one is the Committee for Spatial Development of the German-Polish Intergovernmental Commission. The second one is the recently established German-Czech Intergovernmental Working Group for Spatial Development, which is presented in more detail in one of our case studies below. Furthermore, there are several co-ordination units at the regional level, for example between German states (Länder) and Polish voivodeships.

Due to different paths of transition the political and administrative systems in the three countries differ considerably. While the current administrative system has existed in western Germany since 1949 and in eastern Germany basically since 1990, with changes in 1994, the structures in Poland and the Czech Republic are significantly younger. The current administrative system in Poland with 16 voivodeships has only been in existence since 1999, and the one in the Czech Republic with 14 districts (kraj) since 2003. While there are six political-administrational levels in Saxony and Bavaria, the Polish system comprises only four and in the Czech system there are three such levels. This fact requires the willingness of the involved actors to co-operate with partners who do not always match their own hierarchical rank.[36]

The repercussions of the administrational reforms in Poland and the Czech Republic still cause a labour turnover in the administrations and hinder the process of cross-border trust-building, which is important since trust- and network-building are based on personal contacts between the co-operating partners. One can observe uncertainties among the administrative staff that can be traced back to frequently changing and hardly established responsibilities.[37]

Furthermore, transboundary governance between Germany and Poland or Germany and the Czech Republic is often hampered by cultural differences, prejudices and language barriers. There are hardly any similarities between the Slavic languages Polish and Czech on the one hand, and Germanic languages on the other hand. The conjunction of different languages is not unique, but what is unique here is the low degree of proficiency in the neighbours' languages, especially on the German side of the borders. Difficult historic experiences also contribute to complex overall relationships between the populations of the three countries.[38]

With regard to existing transboundary organisations and institutions, particularly the Euroregions at the communal level are engaged in cross-border co-operation. There are eight Euroregions along the German-Polish and German-Czech borders (see Fig. 1). These are mainly voluntary associations of municipalities. The Euroregions have two-tiered organisational structures with separate offices in each country and separate administrative structures. Besides the activities of the Euroregions, numerous co-operation activities are taking place at the local level. On the one hand, they consist of the co-operation of adjoining cities focusing on

[36] Dendewicz (2001) pp 10-15; Droth et al. (2000) pp 101-104; ÚÚR (2002) pp 37-39
[37] Klein-Hitpaß et al. (2006) pp 9-62
[38] Jajeśniak-Quast/Stokłosa (2000)

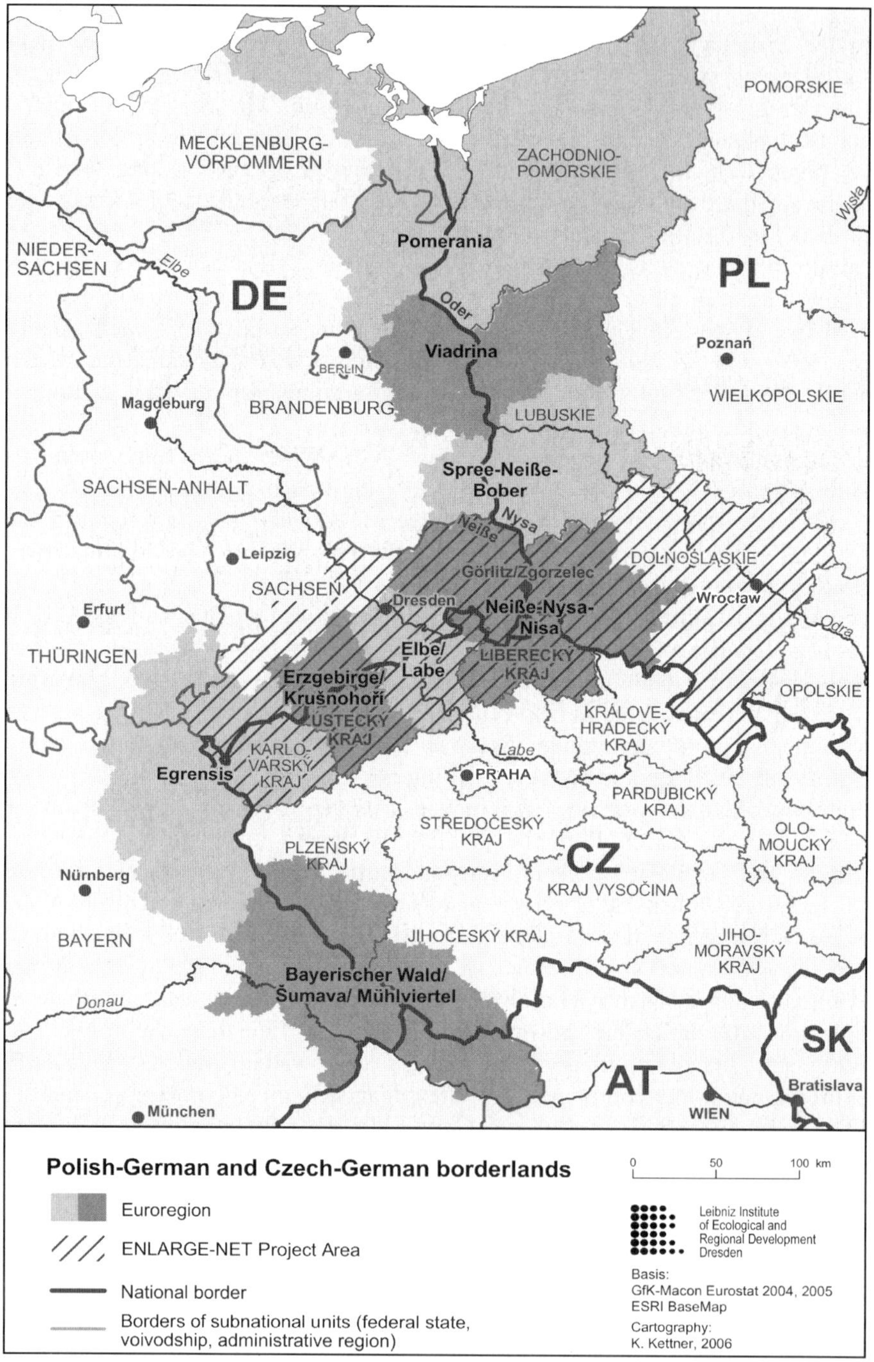

Fig.1: Polish-German and Czech-German Borderlands. Design: IOER (2006)

Fig.2: The Old Town of Görlitz on the right side faces derelict land in Zgorzelec on the opposite bank of the river Neiße. Photo: H. Heil, City of Görlitz.

co-ordinated land use planning and/or urban management. On the other hand, some cities are active in town twinning projects over longer distances, for example Dresden and Wrocław or Chemnitz and Ústí nad Labem.[39]

Görlitz/Zgorzelec: transboundary co-ordination at the local level

Due to the reshaping of Poland's and Germany's territories at the end of World War II, the city of Görlitz was divided into two parts: a German city west of the river Neiße and the new Polish city Zgorzelec at the eastern bank of the river. Between 1945 and 1991 the two cities developed in almost complete isolation from each other. The Old Town of Görlitz faces derelict land that is only now being developed, while in the city of Zgorzelec a new centre developed further north after World War II. A bridge rebuilt in 2004 has connected the historic centre again (see Fig. 2).

After Poland introduced visa-free border traffic in 1991, co-operation between the two adjacent cities intensified. In 1993 a partnership treaty was signed. In 1995 the two cities established a joint decision-making body called the Co-ordinating Commission and in 1998 the cities declared themselves as the European City Görlitz/Zgorzelec.

[39] Leibenath/Knippschild (2004) pp 16ff

Between 2002 and 2004 both cities took part in the German federal programme *"Stadt 2030"*, an ideas competition launched by the German Ministry for Education and Research aimed at the development of visions and strategies for cities' future development. Görlitz and Zgorzelec were successful in submitting a project proposal in co-operation with the Leibniz Institute of Ecological and Regional Development in Dresden. The main objective of the project was to deepen co-operation, to reach agreement on integrated city development, and to support trust-building between the involved actors. The project funds allowed the installation of a co-ordination office and scientific consultation during a project period lasting almost two years. The approach was to initiate a development process focusing towards a joint vision by developing and discussing scenarios for both cities. The idea was to start off with the public administrations and open the discussion process to the business community, civil society and citizens at large.

Key elements of the project were five existing (but not well-established) thematic working groups within the two municipalities (City Planning and Housing; Economy, City Marketing and Infrastructure; Culture and Tourism; Youth, Education and Sports; and Health, Social Affairs and Environment) and several conferences with a broader group of participants. The co-ordination office and the scientists were in charge of moderating the working groups, adding thematic input to the discussion process, organising the conferences, processing the results, and evaluating the communication process. Since the Co-ordinating Commission, which tackled the co-ordination of everyday tasks, was already well established, the project *"Stadt 2030"* could instead focus on coordinating vision and strategy development.

What results did the transboundary discussion process generate in the end? The scientists involved have provided scenarios for the common future development of the two cities. These scenarios deal with the entire border region's economy and population, its identity, and its attitudes and development.[40] These topics were identified as being important by the working groups at an earlier point. The scenarios were the basis for all subsequent results. The five working groups developed ideas for future development within their topic. These ideas ranged from a Bridge Park, urban development along the river which would support education, culture and arts, two common libraries, a common museum, and the elaboration of common environmental quality targets.

Moreover, general rules for co-operation were elaborated in a cross-sectional manner. These rules formulate self-imposed demands for future co-operation as a result of the co-operation process: The political legitimacy for transboundary acting has to be secured. Symbolic co-operation has to be replaced by the reaching of joint decisions and by acting together. In a long-term perspective, administrative and communicative structures have to be changed. This includes giving certain issues of

[40] Neumann (2005)

co-operation to new transboundary organisations. The citizens have to be involved intensively since transboundary structures and the joint development of the two cities can only be successful when accepted by the public.[41]

Since all these project ideas mentioned above could not be realised in the near future due to restricted financial means, the question was: What are the priorities of co-operation? Which projects should begin first? The most important and prominent project seemed to be the Bridge Park. An artistic and scientific concept for this project was elaborated within the project, and could be integrated into the application for the European Capital of Culture 2010 for both cities in the year 2006. The second priority was to create a joint economic strategy for Görlitz and Zgorzelec. The third priority was to work out a cross-sectional development strategy for both cities. The second and the third ideas have since been realised within an INTERREG III A project (about a joint economic and development strategy for both cities) conducted by the city of Zgorzelec.[42]

The results of the *Stadt 2030* project have led to modifications being made in the partnership treaty, which is now more detailed and up to date, and concerned with present challenges and opportunities for both cities.[43]

How can the Görlitz/Zgorzelec case study be reviewed in the end regarding transboundary co-ordination of spatial policies at the local level? Although the project *"Stadt 2030"* was successful in initiating a discussion process about a joint vision and strategic orientation, it showed deficits as well. At a certain point the administrative staff of both cities suffered from a lack of political backing for their transboundary acting and decision-making. It is possible that the political elites of both cities were involved into the project too late and in too vague a fashion. Another weak point was the opening of the working groups to non-administrative actors. Involving entrepreneurship, the civil society and citizens was difficult. The problems of administrative and institutional asymmetries between the cities were not too evident. Problems arose concerning environmental protection since this competence is not on the municipal, but on the district level in Poland. Bigger problems were caused by unequal personnel resources within both administrations. While in Görlitz an entire department with dozens of employees is responsible for city planning, this task in Zgorzelec is performed by a single colleague.[44]

The fact that both cities had co-operated with each other for many years had a positive influence on the project. Administrational staff is to some extent able and willing to discuss strategic topics. There is a partnership treaty that ensures the political will to co-operate and the Co-ordinating Commission allows for joint decisions to be made.

---

[41] Książek/Kunert (2004) pp 57-62
[42] Książek/Kunert (2004) pp 63-65
[43] Książek/Kunert (2004) pp 57
[44] Knippschild/Kunert (2005)

Enlarge-Net: transboundary co-ordination at the regional level

Enlarge-Net has been an attempt at transboundary co-ordination between three countries. It was a project which was funded by the European Commission and had the full title: "Cross-border networks between Cities and Regions in Saxony, Lower Silesia and Northern Bohemia in preparation for EU enlargement." Enlarge-Net brought together the four cities of Wrocław, Ústí nad Labem, Chemnitz, and Dresden (the latter being the lead partner in the project), the three regions of the Lower SilesiaVoivodeship ("Województwo Dolnośląskie"), the district of Ústí nad Labem ("Ústecký kraj") and the district of Dresden ("Regierungsbezirk Dresden"). Moreover, there were two public institutes in the group of project partners: the Saxon Communal Education and Training Institute (SKSD) and the Leibniz Institute of Ecological and Regional Development (IOER), both based in Dresden. Altogether there were two Polish, two Czech and five German partners. Furthermore, the three regions composed of the district of Chemnitz ("Regierungsbezirk Chemnitz"), the district of Karlovy Vary ("Karlovarský kraj"), and district of Liberec ("Liberecký kraj") co-operated without being formal partners. Enlarge-Net ran for two years from mid-2003 to mid-2005.

The objective of Enlarge-Net was first of all to bring together administrative staff from local and regional authorities and to offer them a platform for becoming acquainted, for mutual information on administrative structures and procedures, and for exchanging views. The network, however, was not limited to public officials: By integrating actors from civil society who are part of existing cross-border networks, it was intended to establish a so-called network of the networks. The second major objective of the project was to develop joint transboundary analyses, strategies and measures with regard to the effects of EU enlargement. During the course of the project this second aim received growing attention and overshadowed all other concerns.

The structure of the project comprised a steering committee which joined the political representatives of the co-operating organizations. The steering committee was accompanied by a co-ordination group which met more frequently and a project secretariat. The project secretariat functioned as a contact point for all participants as well as for other interested parties. At the heart of the networking process were five thematic working groups. Each working group was formed by 25-35 experts from Poland, the Czech Republic and Germany and held two-day workshops twice a year. In all, there were four workshops per group, or a total of 20 workshops. Moreover, there were additional meetings of smaller subgroups between workshops. The working groups were chaired by multi-national teams of two or three persons each and the discussions were facilitated by IOER staff.[45]

[45] Leibenath/Knippschild (2005b) p 81

Four out of the five working groups covered topics with a spatial relevance:

- Migration and social issues,
- regional planning, infrastructure and transport,
- the environment and flood protection, and
- the economy and tourism.

Especially the group on regional planning identified a broad range of issues which were expected to require some sort of transboundary co-ordination, for example the revitalization of brownfield sites, the improvement of cross-border transport links, and the mutual adjustment of spatial plans. The project manager and also some of the working group members were very eager to get tangible results in the form of specific follow-up projects and tried to push the groups in this direction. These efforts foundered with only very few modest exceptions, although both the working group members and the facilitators tried hard to define specific activities. What were the reasons? First, one has to mention the harsh differences in language and culture which made real co-operation very difficult. Communication styles differed very much, for example with regard to the importance of formalised, hierarchical flows of information or with regard to the preferred medium of information – regular mail, e-mail, phone or face-to-face. For many participants it was impossible to communicate without an interpreter because there was no common language.

The second major obstacle was the unstable composition of the working groups. Only a small share of actors attended all four consecutive rounds of the workshops. This was due to the usual time constraints, but also to staff rotations, especially in participating Polish and Czech organisations. Another reason was that some of the working group members realised that Enlarge-Net would not meet their needs. Whatever the individual reasons might have been: the lack of stability made it impossible in many cases to build up personal trust and to reach specific agreements.

The project area was very large: It had an east-west extension of approximately 400 kilometres and a north-south extension of approximately 150 kilometres with overall travelling times between the partners of up to six hours. This not only made face-to-face contacts very costly, but also hampered the identification of real joint challenges or tasks for transboundary co-ordination: There were virtually no spatial concerns that affected the entire project area from Wrocław to Chemnitz and fell within the competencies of the participating agencies and organisations. The issues at stake either touched only a small part of the territory, for example the stabilisation of bogs on top of a ridge in the Ore Mountains, or would have required the involvement of state or national authorities, for example the preparation of a transboundary development concept.

What were the achievements of Enlarge-Net with regard to the transboundary co-ordination of spatial policies? Although participants were somewhat disappointed about the lack of tangible results, they did not judge it as a failure. Instead, two surveys on the basis of semi-standardized questionnaires filled in by all participants of the second and the fourth rounds of workshops revealed that the project was highly valued because of the contacts that had been established and because of the information that had been obtained. One has to be aware that more than 40 % of the participants did not have any previous professional contacts in these neighbouring countries. And even those who had previous contacts didn't know much about their neighbours' institutions, perceptions and interests. Thus, Enlarge-Net contributed significantly to the improvement of transboundary institutional capacity in the project region. This is reflected nicely by the results of the second survey among the 73 persons who participated in the fourth round of workshops. 93 % answered that trust had increased between the participants, 98 % answered that there was a better mutual understanding of concerns and interests, 86 % answered that they felt more secure and comfortable in cross-border co-operation, and 80 % answered that the information they got through Enlarge-Net helped them in doing their daily work.[46] It is somewhat questionable, however, just how accurate the results of the survey are because the answers may have been biased.

German-Czech Working Group for Spatial Development: transboundary co-ordination at the level of national governments

A bilateral intergovernmental co-ordinating body for spatial planning and development between Germany and the Czech Republic did not come into existence until 2006. The simplest explanation would be that neither side took the initiative. A closer look reveals that some of the most pressing transboundary spatial issues were already covered by the German-Czech Environmental Commission established in 1996 and the International Commission for the Protection of the river Elbe/Labe which even dates back to 1990. Another cause might be the reform of the public administration in the Czech Republic in 2000 which resulted in enduring fluctuations of staff and shifting responsibilities.

However, in 2004 officials from the German Federal Ministry for Transport, Building and Housing (BMVBW)[47] established new contacts to the Czech Ministry for Regional Development (MMR) and negotiated the initiation of a joint working group for spatial development. In 2005 the BMVBW contracted the Leibniz Institute of Ecological and Regional Development (IOER) in order to elaborate a preparatory study in co-operation with the Czech Institute for Spatial Development (ÚÚR). The terms of reference for the study included – among other things – identifying key issues of transboundary spatial development and proposing a spatial vision for the development of the extended Czech-German border region which was defined as the area between Prague, Munich, Berlin and the Polish city of Wrocław (see Fig. 3). To this end, the authors analysed a variety of written material and conducted a phone survey among all spatial planning agencies on both sides of the border as well

[46] Leibenath/Knippschild (2005a) p 14 and p 19

[47] It has been renamed the German Federal Ministry for Transport, Building and Urban Development (BMVBS)

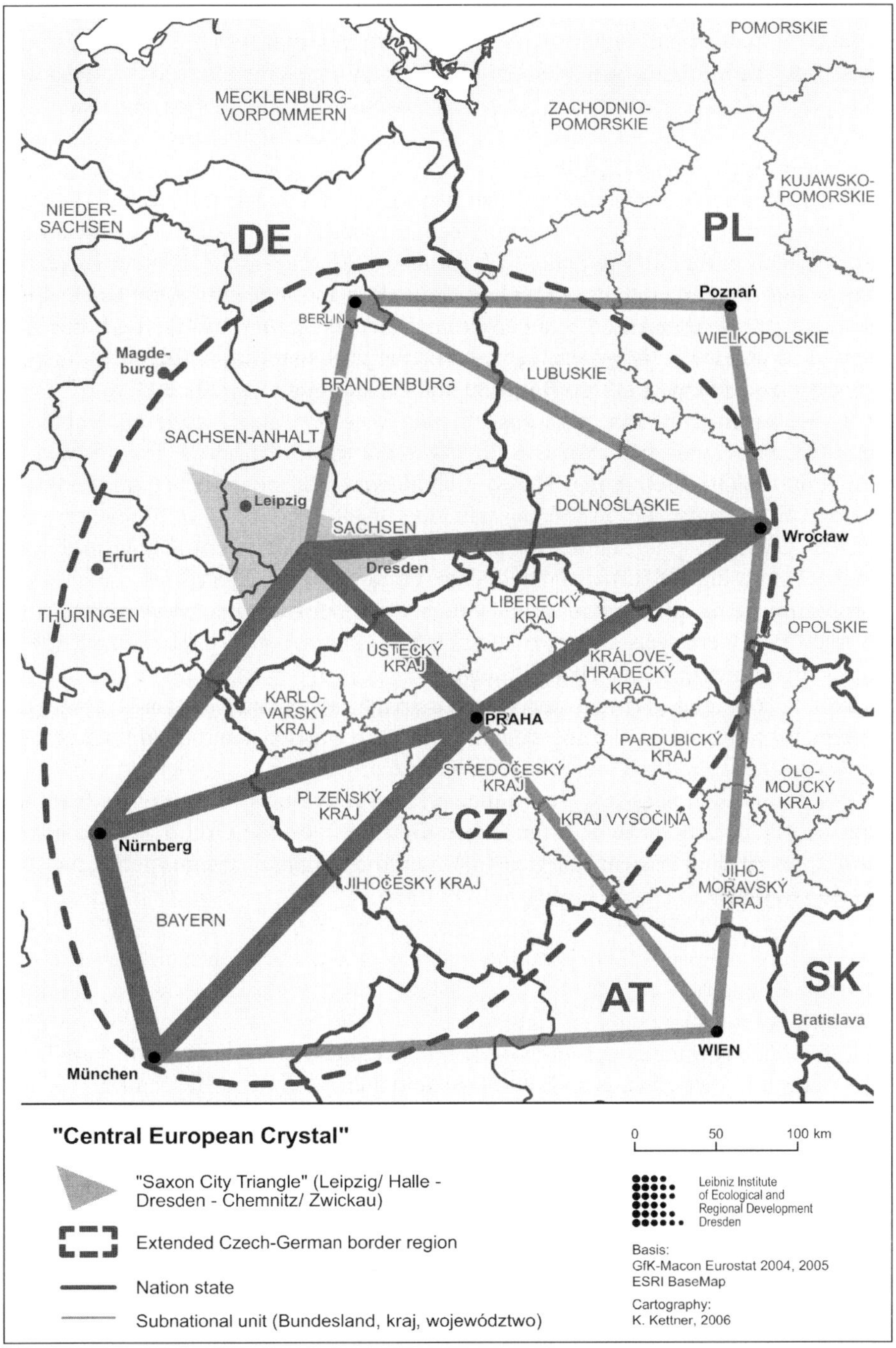

Fig.3: Extended Czech-German border area and the proposed vision of the "Central European Crystal". Design: IOER (2006)

as holding personal in-depth interviews with representatives of the two national ministries. Additionally, a public seminar for planning officials from regional and national levels was organized in order to discuss the findings of the study and to formulate recommendations for the working group which convened its initial meeting in April 2006 in Berlin.

The perception of spatial issues that needed transboundary co-ordination at the intergovernmental level differed significantly among Czech planning officials on the one hand and their German counterparts on the other hand. By and large, the Czechs did not perceive any urgent co-ordination requirements at all and mainly mentioned the potential for a better co-ordination of economic affairs, for example in the tourism industry. They merely expressed the wish that certain regional transport connections between southern Bohemia and Bavaria receive more attention.

By contrast, the German planning officials were very much concerned about the construction of wind power plants directly adjacent to the border and located on a ridge in the Ore Mountains. They anticipated an adverse influence on the visual quality of the landscape on the German side of the border and felt they were not offered sufficient opportunities for participation in the decision-making processes on the Czech side. This also applied to large commercial centres which were supposedly being planned in some Czech border towns. A closer investigation revealed that there were just two such centres planned, with uncertain prospects regarding their actual construction, however.

Both Czech and German survey participants articulated concerns about the asymmetry of national planning systems and were worried about insufficient cross-border information flows and related linguistic difficulties.

At the seminar and especially in the personal interviews with officials from the ministries it became clear that the Czech-German Working Group is supposed to fulfil many of the classical functions an intergovernmental commission for spatial development has:

- Prepare and disseminate information on spatial planning systems and terminologies,
- analyse transboundary spatial issues,
- elaborate recommendations for the transboundary co-ordination of spatial plans and individual projects with potential cross-border effects, and
- support or even initiate co-operation on transboundary spatial visions and concepts.

Even though the Czech-German Working Group for Spatial Development was only founded recently, some remarks can be made with respect to the intended mode of transboundary co-ordination. The leading actors from the BMVBS explicitly do not want to have a commission in the traditional sense with a strong involvement of the ministries for foreign affairs, formalised procedures, and so on, because of negative experience with similar organisations in which the BMVBS was involved.

Instead they desire a flexible co-operative structure that can function as a facilitator or catalyst for spatial co-ordination between actors from different levels of policy-making, for example spatial planners from the regions, business representatives, and non-governmental organisations. The main actors from both the Czech and the German sides are well aware that their influence is very limited in their home administrations, because any time "hard" issues like the environment, transport, or the like are affected, other more specialized sectors of their governments are in charge. Thus, one way they can achieve spatial co-ordination is to become a 'linchpin' that links different sectors and policies that would otherwise be separated. This means that the working group chiefly will be an open platform with a core group of members, but which at the same time will be open in terms of issues and also with regard to other actors that might be called in if needed.

Another instrument at the working group's disposal is the ability to influence mental frames via spatial visions. That is why they included the task of drafting a spatial image for the extended Czech-German border area in the terms of reference for the preparatory study. The study's authors eventually came up with the idea of the "Central European Crystal" (CEC).[48] This image refers to the century-old tradition of glass and china production in the larger region and symbolises its permeability, but also its multifaceted potentials and strengths. Officials from the ministries, once confronted with the CEC, welcomed it strongly. At the charting session of the working group, high-level representatives from the German BMVBS and the Czech MMR even unanimously announced the contracting of another, more detailed study aimed at launching a visioning process for the Czech-German border area based on the CEC.

Finally, the initiators of the working group want to form a coalition which can perform more effective lobbying at the European level. The actors envisage that common spatial concerns of both countries will gain more momentum if articulated with one voice. Furthermore, they agreed to co-operate closer within the framework of the new Objective 3 of European cohesion policy and to co-ordinate their ideas for strategic transnational projects at an early stage.

9.3 Conclusions

Scale matters, but how?

In this paper the transboundary co-ordination of spatial policies has been presented as an approach to territorial cohesion. The underlying assumption is that the transboundary co-ordination of spatial policies can contribute to the reduction of territorial imbalances – both between immediately adjoining regions on different sides of a national border and between centres and peripheries of the European continent as a whole. Starting from theoretical considerations on transboundary

[48] Leibenath et al. (2006)

governance, the three case studies from Germany's eastern borders illuminated recent attempts to reach the transboundary co-ordination of spatial policies. With the following conclusions we want to link the empirical evidence with our initial theoretical considerations. The conclusions are to be understood as preliminary hypotheses which shall incite further research.

With regard to structural framework conditions it can be said that pronounced economic and social imbalances do indeed exist along the Polish-German and Czech-German borders. If, however, Poland's and the Czech Republic's western regions can maintain their current pace of economic growth, the differences will be levelled out within the next few decades – at least in some of the respective border areas. This trend is fuelled by the integration of central European economies into the single European market and by making use of wage and tax differences which favour the western parts of Poland and the Czech Republic.

The cases have shown that actors from all three countries perceive a need for a stronger transboundary co-ordination of spatial policies which, at the same time, remains a challenging and demanding business. While formal co-ordination mechanisms are in their infancy and very rarely practiced here – for example in the frame of strategic environmental assessments or project-related environmental impact assessments – the emphasis in all three cases was on the co-ordination of urban and spatial development issues instead of legally binding regulatory plans or measures.

The cases highlight the fact that multi-level governance in European transboundary governance is widespread. This can be attributed to the strong interrelations and dependencies of different levels of policy-making. In the case of Görlitz/Zgorzelec the analysed project could only be launched thanks to funding from the German national government. In the case of Enlarge-Net, funds from the European Commission were the first and main *raison d'être*. The scope of Enlarge-Net itself was also very comprehensive and encompassed actors from local and various regional levels. And finally, the case study on the German-Czech Working Group for Spatial Development revealed that even national ministries responsible for spatial development have realised that they are only a small part of the bigger game at the European level (new Objective 3 of the cohesion policy, etc.) and that they depend on various subnational networks and actors. Hence, the case studies underpin the hypothesis that transboundary governance in Europe is characterised by multi-level structures, but also by a dominance of public actors and organisations and – at least with regard to the Görlitz/Zgorzelec and Enlarge-Net cases – by a strong reliance on public funding.[49] However, none of the three cases can be regarded as particularly successful in co-ordinating spatial policies across boundaries and simultaneously across political-administrative levels.

[49] see also Gualini (2003) p 48; Scott (1999) pp 609f

Territorial cohesion policies in transboundary contexts are prone to run into a complexity dilemma because, if taken seriously, they even require three-tiered approaches, i.e. co-ordination across boundaries, across levels and across subject-related sectors of policy-making. It is hard to envisage how such co-ordination could be achieved anyway, but the analysed cases were definitely far from such ambitions. And yet, transboundary co-ordination at the local level between the twin cities Görlitz and Zgorzelec seemed to come closest to this ideal because the political competencies and resources of administrations on both sides are relatively similar in comparison to Enlarge-Net. What is more, the issues to be co-ordinated were easier to define and fitted better into the range of competencies of the two city governments.

The issue of scale is important with regard to capacity-building. As has been shown in general terms above, transboundary policy co-ordination requires some institutional capacity or "institutional thickness"[50] as well as social capital in order to allow for effective co-ordination. The establishment of these qualities is very time consuming and can even founder in case of frequent personal turnovers and unfavourable framework conditions. These caveats apply especially to the intermediate level between local and national governments, because it is here that institutional asymmetries come to the fore in the most articulated way, as the Enlarge-Net case study demonstrated. By contrast, the resources and competencies of local and national governments match better and guarantee more stable relationships even if individual staff members change. This, in turn, leads to a longer "shadow of the future" and to a higher visibility of transactions for other relevant actors. Both factors tend to reduce the risk of opportunistic behaviour.[51] Consequently, it can be concluded that it seems to be easier to move to more advanced forms of co-ordination which are consistent with a narrow understanding of transnationality:

- The shorter the distance between partners is,
- the better the partners' competencies and resources are defined and communicated, and the higher the degree of symmetry between them is, and finally
- the longer and the better individual representatives from both sides know each other.

Transboundary governance at the local level between border cities like Görlitz and Zgorzelec is also an example of the coincidence of functional and territorial governance: The spaces of flows and places overlap neatly because the manifold functional interdependencies between such cities are to a large extent in line with the cities' formal competencies and jurisdictions. The regional level represents the exact opposite, as for instance the formal competencies of a Polish voivodeship,

[50] Gibbs et al. (2001) pp 103f
[51] Scharpf (1997) p 137

a Czech kraj and a German Regierungsbezirk rarely match. This is very likely to lead to constant confusion. The Euroregions have bypassed this obstacle because they are composed of municipalities and not of regional units. As the boundaries of Euroregions have been defined more or less arbitrarily and seldom on the basis of functional transboundary relationships, however, there is much space left for institutional ‘designers’. Enlarge-Net was an example of experimental institution-building which failed, though. Neither a functionally consistent nor a territorially defined architecture of governance could be established because the formal and territorial competencies of the partners didn't match and because the size of the project area did not reflect the few existing large-scale functional interdependencies.

This leads yet to another conclusion concerning the role of project-based external funding. Mainly because of the profound language barriers, but also partly because of long travel distances and the poor condition of public budgets, transboundary co-ordination at the Polish-German or Czech-German borders is largely dependent on external financial assistance, which is usually granted through projects that run not longer than two or three years. This applies to the cases of Görlitz/Zgorzelec and Enlarge-Net. There are major differences, however: In the border cities Görlitz and Zgorzelec, which already had a track record of co-operation and where functional and territorial governance coincide, the *Stadt 2030* project was but one element in an ongoing transboundary governance venture. By contrast, Enlarge-Net fell into neglect after its termination for obvious reasons.

However – and this has also been made evident by the case study – even projects like Enlarge-Net can have valuable functions in terms of “planning as communication” and contribute to the overall establishment of institutional capacity across borders. Although Enlarge-Net has not developed into a forum for a longer-lasting exchange or co-operation, those who participated did benefit significantly with regard to information and experience. This will facilitate any future transboundary co-ordination of spatial policies in other constellations.

Recommendations for policy makers

The theoretical considerations in combination with the three case studies do not only allow for academic conclusions but for practical recommendations geared towards policy makers on the national and European scenes as well. Some of the following proposals have already been formulated in the European Spatial Development Perspective,[52] however, against the backdrop of the findings presented above these ideas can be specified and be made more concrete. The political recommendations might also be of interest for the ongoing European discussions on a Territorial Agenda for the EU.

The first recommendation is to facilitate and to improve the exchange of experiences among different European border regions and among actors from different political and administrative levels. The case studies have mainly highlighted

[52] ESDP; see CSD (1999)

the difficulties associated with the transboundary co-ordination of spatial policies. This may have its causes in the specific nature of the cases, for example their location at a new internal EU border, the low overall degree of transboundary interactions, or the insufficient institutional thickness. Other border regions may do better or follow more innovative approaches. The dissemination of information on experience in different regions of Europe – and maybe even other continents – should be intensified. There are already some fora where such issues are discussed, such as the Association of Borderland Studies (ABS), the Association of European Border Regions (AEBR), and the EU INTERACT programme. The shortcomings of these initiatives or organisations is that they either focus only on scientific questions or only on cross-border co-operation in general, or on a specific funding programme like INTERREG III. What is needed is a permanent platform that integrates both practical experience and theoretical reflections on cross-border spatial planning and the transboundary co-ordination of spatial policies, i.e., a permanent unit for the monitoring and dissemination of transboundary spatial planning issues.

With the European Grouping of Territorial Co-operation,[53] the EU has established an instrument which facilitates the institutionalisation of cross-border co-operation. This novel type of organisation, however, will not necessarily ensure a better transboundary co-ordination of spatial policies. An approach that would be more effective – and this is our second recommendation – is an obligation to assess the territorial impacts of national policies on neighbouring countries. A similar requirement is included in the EU directive on strategic environmental assessments[54] which naturally relates only to environmental concerns, though. The ESDP has already called for cross-border territorial impact assessments (TIA) and defined a TIA as a procedure of consulting the parties who are affected by a measure with considerable spatial impact.[55] Yet, so far there are few cases where contiguous countries have agreed on explicit procedures about how to organise cross-border consultations on plans or projects which could have a major spatial impact on the neighbour's territory.

The third recommendation relates to the elaboration of transboundary spatial concepts. There are already many examples of such documents, for example the Atlantic Spatial Development Perspective.[56] Although these documents are usually entirely informal and non-binding, they can help to establish a common understanding of cross-border problems and priorities for action. They can stimulate processes of re-framing among the actors involved and thus contribute to creating a favourable environment for transnationality in the narrow sense. With regard to borders between new EU member states where only few such concepts exist, the EU and respective national governments should engage in the co-financing of transboundary spatial development concepts because most regional and local governments will be unable

[53] EGTC, see OJEC (2006 /L 210/19); see also the article of Janssen in this volume.
[54] OJEC (2001/L 197/30)
[55] CSD (1999) p 40 and p 45
[56] CPMRE (2005)

to allocate sufficient funding to this end. In this way important contributions to territorial cohesion, to the transboundary co-ordination of spatial policies, and to the competitiveness of border regions could be made.

References

Blatter J (2004) From 'Spaces of Place' to 'Spaces of Flows'? Territorial and Functional Governance in Cross-border Regions in Europe and North-America. In: *International Journal of Urban and Regional Research*, Issue 28(3), pp 530-548

Brunet-Jailly E (2004) Toward a Model of Border Studies: What Do We Learn from the Study of the Canadian-American Border? In: *Journal of Borderlands Studies*, Issue 19(1), pp 1-12

Castells M (1996), The rise of the network society, Oxford, Blackwell

Christiansen T, Jørgensen KE (2000) Transnational Governance 'above' and 'below' the State: The Changing Nature of Borders in the New Europe. In: *Regional and Federal Studies*, Issue 10, pp 62-77

CPMRE (Conference of Peripheral Maritime Regions of Europe. Forward Studies Unit for the Maritime Peripheries) (2005) Atlantic Spatial Development Perspective. Document to be Discussed During the Steering Committee Meeting of 14th June 2005. Available online: http://arcatlantique.org/pdf/en_sdea_synthese_final_agcaa.pdf#search=%22atlantic%20spatial%20development%20perspective%22, visited on 23.08.2006

CSD (Committee on Spatial Development) (1999) ESDP European Spatial Development Perspective. Towards Balanced and Sustainable Development of the Territory of the European Union. Available online: http://europa.eu.int/comm/regional_policy/sources/docoffic/official/reports/pdf/sum_en.pdf, visited on 10.04.2006

de Vries J (2004) Venturing into Unknown Territory: The Making of the Second Benelux Structure Scheme. Paper presented at the 2004 AESOP congress in Grenoble, France

Dendewicz S (2001) Bestandteile des Regionalplanungsystems und die Raumplanung in Polen [Elements of the Regional Planning Systems in Poland]. In: *Bulletin des niederschlesisch-sächsischen Grenzgebietes*, Issue 1/2001, pp 7-16

DG Regio (European Commission, Directorate-General Regional Policy) (2004) Interim Territorial Cohesion Report (Preliminary results of ESPON and EU Commission studies). Available online: http://ec.europa.eu/regional_policy/sources/docoffic/official/reports/coheter/coheter_en.pdf, visited on 20.06.2006

Droth A, Grimm FD, Haase A (2000) Polen aktuell [Current Poland], Institut für Länderkunde Leipzig

ECTP (European Council of Town Planners), CSD (Committee on Spatial Development) (2001) Proceedings of the one-day conference on territorial impact assessment. Available online: http://www.ceu-ectp.org/inc/cgi/cp/cp20030515.pdf#search=%22louvain-la-neuve%20territorial%20impact%20assessment%22, visited on 23.08.2006

EU-COM (European Commission) (2001) Unity, solidarity, diversity for Europe, its people and its territory. Second report on economic and social cohesion. Available online: http://ec.europa.eu/regional_policy/sources/docoffic/official/reports/contentpdf_en.htm, visited on 30.06.2006

EU-COM (European Commission) (2003) Structural policies and European territories. Competitiveness, sustainable development and cohesion in Europe. From Lisbon to Gothenburg. Available online: http://ec.europa.eu/regional_policy/sources/docgener/presenta/comp/comp_en.pdf, visited on 20.06.2006

EU-COM (European Commission) (2004) A new partnership for cohesion. Convergence, competitiveness, cooperation. Available online: http://ec.europa.eu/regional_policy/sources/docoffic/official/reports/cohesion3/cohesion3_en.htm, visited on 20.06.2006

Faludi A (2003) Territorial Cohesion: Old (French) Wine in New Bottles? Available online: http://www.ru.nl/gap/papers/gapwp03-6.pdf, visited on 27.06.2005

Fürst D (2003) Steuerung auf regionaler Ebene versus Regional Governance [Policy-making at Regional Level versus Regional Governance]. In: *Informationen zur Raumentwicklung*, Issue 8/9.2003, pp 441-450

Gibbs DC, Jonas AEG, Reimer S, Spooner DJ (2001) Governance, institutional capacity and partnerships in local economic development: theoretical issues and empirical evidence from the Humber Sub-region. In: Transactions of the Institute of British Geographers, Issue 26, pp 103-119

Gualini E (2003) Cross-border Governance: Inventing Regions in a Trans-national Multi-level Polity, DISP 152, pp 43-52

Gualini E (2005) Europeanization and Governance Rescaling: “Territorial Cohesion” as Opportunity for “Re-politicizing” European Spatial Policy and Planning? Available online: http://aesop2005.scix.net/cgi-bin/papers/Show?_id=340, visited on 19.08.2005

Gustedt E (2005) The Territorial Cohesion of Europe – from a Vague Notion to a Clear Idea. Available online: http://aesop2005.scix.net/cgi-bin/papers/Show?_id=467, visited on 19.08.2005

Healey P (1997) Collaborative Planning, London

Healey P (1998) Building institutional capacity through collaborative approaches to urban planning. In: Environment and Planning A, Issue 30, pp 1531-1546

Healey P (2006) Transforming governance: Challenges of institutional adaptation and a new politics of space. In: *European Planning Studies*, Issue 14(3), pp 299-320

Jajeśniak-Quast D, Stokłosa K (2000) Geteilte Städte an Oder und Neiße. Frankfurt (Oder) – Słubice, Guben – Gubin und Görlitz – Zgorzelec 1945-1995 [Devided Cities on Odra and Nysa rivers. Frankfurt (Oder) – Słubice, Guben – Gubin und Görlitz – Zgorzelec 1945-1995], *Frankfurter Studien zur Grenzregion*, Band 5, Berlin, Berlin Verlag

Jänicke M (2002) The Political System's Capacity for Environmental Policy: The Framework for Comparison. In: Weidner H, Jänicke M (eds.) Capacity Building in National Environmental Policy. Berlin, Springer, pp 1-18

Klein-Hitpaß K, Leibenath M, Knippschild R (2006) Vertrauen in grenzüberschreitenden Akteursnetzwerken. Erkenntnisse aus dem deutsch-polnisch-tschechischen Kooperationsprojekt ENLARGE-NET [Trust in Cross-border Networks of Actors. Findings from the German-Polish-Czech Cooperation Project ENLARGE-NET]. disp 164, pp 59-70. Available online: http://www.nsl.ethz.ch/index.php/de/content/download/1261/7608/file/, visited on 27.06.2006

Knippschild R, Kunert M (2005) Kommunikation und Beteiligung bei der grenzüberschreitenden Leitbildentwicklung in Görlitz/Zgorzelec [Communication and Participation for Cross-border Visioning in Görlitz/Zgorzelec]. In: Friedrich K, Knippschild R, Kunert M, Meyer-Künzel M, Neumann I (eds.) Zwei Grenzstädte wachsen zusammen – Im Zukunftsdialog zur Europastadt Görlitz/Zgorzelec [Two Border Cities Growing Together – With a Dialogue about the Future towards the European City Görlitz/Zgorzelec]. München, Oekom, pp 65-80

Książek K, Kunert M (eds.) (2004) Stadt 2030 – Gemeinsames Leitbild für die Europastadt Görlitz/Zgorzelec. Grenzüberschreitende Leitbildentwicklung in einer Stadt zweier Nationen [Joint Vision for the Europe City Görlitz/Zgorzelec. Cross-border visioning in a City of Two Nations]. Dresden, Thelem

Leibenath M, Knippschild R (2004) ENLARGE-NET – Dritter Evaluierungsbericht [Third Evaluation Report]. Available online: http://enlarge-net.dresden.de/enlarge-net/de/a1/b1/PDF/Evaluierung3.pdf, visited on 22.06.2006

Leibenath M, Knippschild R (2005a) ENLARGE-NET – Vierter Evaluierungsbericht [Fourth Evaluation Report]. Available online: http://enlarge-net.dresden.de/enlarge-net/de/a1/b1/PDF/Evaluierung4.pdf, visited on 22.06.2006

Leibenath M, Knippschild R (2005b) Systemic Evaluation of Cross-Border Networks of Actors: Experience with a German-Polish-Czech Cooperation Project. In: *Journal of Borderlands Studies*, Issue 20, pp 73-90

Leibenath M, Knippschild R, Hahn A, Simon D, Kyselka I, Marvart J, Polešáková M, Veselý R (2006) Synthese gemeinsamer deutsch-tschechischer Raumordnungsvorhaben zur Stärkung der territorialen Kohäsion Zwischenbericht [Interim Report on the Project "Synopsis of Common German-Czech Spatial Development Projects as a Contribution to Territorial Cohesion"]. Available online: http://www.ioer.de/ioer_projekte/PDF/182 Zw-Bericht Deutsch.pdf, visited on 29.06.2006

Luxembourg Presidency of the Council of the European Union (2005) Scoping document and summary of political messages for an assessment of the Territorial State and Perspectives of the European Union towards a stronger European territorial cohesion in the light of the Lisbon and Gothenburg ambitions. Available online: http://www.eu2005.lu/en/actualites/documents_travail/2005/05/20regio/Min_DOC_1_fin.pdf, visited on 20.06.2006

Neumann I (ed.) (2005) Die Zukunft des deutsch-polnischen Grenzraums gestalten. Szenarien für die Europastadt Görlitz/Zgorzelec im Zuge der EU-Erweiterung [Designing the Future of the German-Polish Borderland. Scenarios for the Europe City Görlitz/Zgorzelec within the Process of EU Enlargement]. Dresden, Thelem

Scharpf FW (1997) Games Real Actors Play. Actor-Centred Institutionalism in Policy Research, Boulder, Westview

Schroll-Machl S, Nový I (2003) Perfekt geplant oder genial improvisiert? Kulturunterschiede in der deutsch-tschechischen Zusammenarbeit [Perfect Planning or Ingenious Improvisation? Cultural Differences in German-Czech Collaboration]. München, Hampp

Scott JW (1999) European and North American contexts for Cross-border Regionalism. In: Regional Studies, Issue 33.7, pp 605-617

Scott JW (2002) Transboundary Governance in the Baltic Sea Region. Emerging Patterns, Preliminary Results. In: *Regional and Federal Studies*, Issue 12(4), pp 135-153

SMI (Saxon Ministry of the Interior) (2004) Rothenburger Erklärung zur grenzüberschreitenden Regionalentwicklung Rothenburg [Declaration for Cross-Border Regional Development]. Available online: http://wo5app1.saxony.de/app/WebObjects/mspublic.woa/wa/componentDetails?id=10422, visited on 02.11.2004

ÚÚR (Ústav Územniího Rozvoje) (ed.) (2002) Town and Country Planning in the Czech Republic, Praha, ÚÚR

Zonneveld W (2005) Expansive Spatial Planning: The New European Transnational Spatial Visions. In: *European Planning Studies*, Issue 13, pp 137-155

Legal Acts

BGBl (Bundesgesetzblatt [Federal Law Gazette]) 1949, III 100-1, Basic Constitutional Law for the Federal Republic of Germany from 23 May 1949, most recently amended by the amending law dated 26 July 2002 (BGBl I, 2863). Available online: http://www.bundesregierung.de/static/pdf/GG_engl_Stand_26_07_02.pdf, visited on 17.10.2005

BGBl (Bundesgesetzblatt [Federal Law Gazette]) 1992, II 462, Deutsch-Tschechischer Staatsvertrag über gute Nachbarschaft von 1992 [German-Czech treaty on good neighbourhood of the year 1992]

BGBl (Bundesgesetzblatt [Federal Law Gazette]) 1991, II 1315, Deutsch-Polnischer Staatsvertrag über gute Nachbarschaft und freundschaftliche Zusammenarbeit von 1991 [German-Czech treaty on good neighbourhood of the year 1991]

BGBl (= Bundesgesetzblatt [Federal Law Gazette]) 1997 II, 1159, Agreement of Karlsruhe of the year 1997

OJEC (Official Journal of the European Union) 2000/L 327/1, Directive 2000/60/EC of the European Parliament and of the Council of 23 October 2000 establishing a framework for Community action in the field of water policy

OJEC (Official Journal of the European Union) 2001/L 197/30, Directive 2001/42/EC of the European Parliament and of the Council of 27 June 2001 on the assessment of the effects of certain plans and programmes on the environment

OJEC (Official Journal of the European Union) 1997/C 340/35, Consolidated Version of the Treaty Establishing the European Community

OJEC (Official Journal of the European Union) 2004/C 310/1, Treaty Establishing a Constitution for Europe

OJEC (Official Journal of the European Union) 2006/L 210/19, Regulation (EC) No 1082/2006 of the European Parliament and of the Council of 5 July 2006 on a European grouping of territorial co-operation (EGTC)

10 Spatial Development Strategies in the Enlarged EU: ESPON Project 1.2.3 on Spatially Relevant Aspects of the Information Society

Kristine Müller, Sabine Zillmer

Introduction

The project presented in this paper has been part of the ESPON 2006 Programme[1] and was financed under the Community Initiative INTERREG III. This programme represents an international network of scientists, politicians and other stakeholders working within the area of European Spatial Development. Among other things, this programme aims at improving the understanding of European territorial diversity and development, and supports the development of suitable instruments and institutions for an enhanced perception and application of ESDP[2] options[3]. Within this framework the ESPON project 1.2.3 – in which the Institute for Regional Development and Structural Planning (IRS) collaborated as a project partner – aimed at the identification of spatially relevant aspects of the information society. On the basis of an analysis of corresponding structures, processes, and policy actions already undertaken, the project developed policy recommendations aimed at supporting the development of an information society in Europe.

Consequently, at the general level the analysis of ESPON project 1.2.3 is strongly linked to the EU objective of territorial cohesion, which supplements the objectives of social and economic cohesion formulated at an earlier date by the European Commission[4]. More specifically, the project is also related to the so-called Lisbon process[5], which aims to develop the EU into the most competitive and dynamic knowledge-based economy in the world. In a wider definition of the term 'information society', such a knowledge-based economy can be considered to be the basis for a society in which knowledge and information are fundamental. Defined this way, an information society is perceived as a learning, innovative, and creative society rather than a society simply making intensive use of telecommunication infrastructures[6].

Thus, in the case of a relatively broad definition, as it has been used for the purpose of ESPON project 1.2.3, an information society has several dimensions which need to be considered. In this kind of setting, the knowledge-based economy is an element of the information society. This, in turn, incorporates the different

---

[1] ESPON – European Spatial Planning Observation Network (see www.espon.eu)
[2] ESDP – European Spatial Development Perspective
[3] European Commission (1999)
[4] ibid.
[5] European Union (2000)
[6] c.f. EUROREG; IRS et al (2006)

aspects of knowledge and information in economic and social life within three different dimensions: technological, economic, and social. The technological dimension, above all, involves the spread of the availability and use of information and communication technologies (ICT). The economic dimension is related to the development of respective economic sectors, including for example, R&D and knowledge intensive services, their performance and their innovation intensities. Finally, the social dimension, above all, describes the role of education and of ICT use in citizens' lives.

These dimensions have served to facilitate the handling of the so far rather vague understanding of the information society, in order to be able to focus and structure the search for appropriate indicators as well as their analysis. To gain a comprehensive understanding of the information society's territorial structures and development, this analysis has been applied to the overall ESPON space[7] as well as – at more detail – to selected national and regional territories.

10.1 Territorial Characteristics of the Information Society in Europe

International sources of analyses at the national level, such as the eEurope[8] and the Networked Readiness Indexes[9], clearly illustrate the considerable spatial differences in the progress of achieving an information society's attributes. North-south and east-west divides occur as the most outstanding divisions. Northern European countries tend to have a more developed information society as compared to southern or eastern European countries. The east-west divide is largely due to the slowness of development of an information society in the new member states of Eastern Europe up to now.

In addition, in order to gain deeper insight into European-wide information society development, a synthetic territorial information society index at the regional level (NUTS 2) has been constructed.[10] Besides the mostly technology oriented indicators applied in the above-mentioned international indexes, this synthetic index was compiled under consideration of the other dimensions of the information society as well. Due to data limitations the social dimension was, however, considered to a lesser extent than the economic dimension. As is pointed out by the map, in the majority of countries, the aforementioned national disparities occur simultaneously with considerable regional disparities.

---

[7] Most projects in the ESPON 2006 Programme analysed the EU 25 countries plus candidate countries Bulgaria and Romania, as well as Norway and Switzerland, as countries participating in the ESPON Programme. The entire group of countries is called the ESPON space.

[8] European Commission, eEurope (2005)

[9] c.f. WEFORUM (2004)

[10] EUROREG, IRS et al (2006)

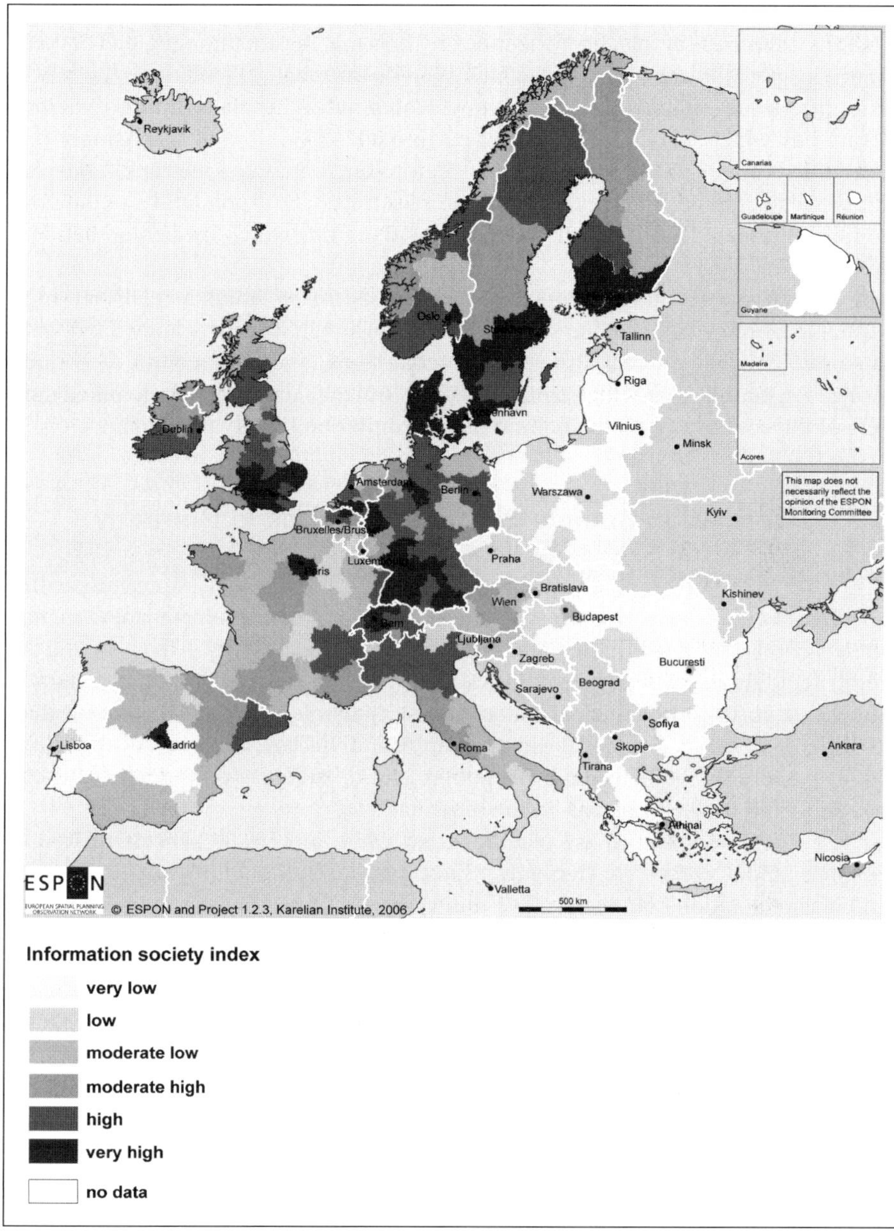

Map 1: The ESPON 1.2.3 Synthetic Information Society Index. Source: EUROREG et al. (2006): Identification of Spatially Relevant Aspects of the Information Society – Final Report ESPON Project 1.2.3, p 14

In Nordic countries for instance, the index reveals that the southern regions of these countries especially contribute to the high average information society performance, while their more peripheral regions show scores no higher than in many regions in countries which have a much poorer performance overall. Not surprisingly this analysis also points out that in particular regions encompassing metropolitan areas rank better than most other regions in these countries. Correspondingly, in the non-Nordic countries it is also usually the peripheral regions which lag behind national average index scores.

In comparison to other European countries, Germany holds a relatively high position in average information society indexes. Though lagging somewhat behind the Nordic countries, Great Britain, the Netherlands, and Switzerland, it is quite advanced in comparison with France, Belgium, Ireland, and Italy, etc., to name only a few of the countries with medium to high information society performance.

10.2 Information Society Status and Trends in Germany

Even more than ten years after Germany's reunification, rather strong disparities prevail between eastern and western Germany. Particular problem areas in this context are the continuous high rates of unemployment, especially in the east, and the overall demographic development. In addition, with regard to many socio-economic indicators a north-south division can be observed, although it is less distinctive than the disparities above. Besides these geographical divisions, Germany is also faced with a spatial division between metropolitan and urban areas on the one hand and more rural and peripheral areas on the other hand[11].

These spatial disparities are also apparent when looking at aspects related to the information society in Germany. Since various types of regions exist with particular population compositions, a digital divide which is usually observed for different age and education groups is also reflected in spatial terms. In particular, the peripheral regions in eastern Germany have the least favourable position in terms of development related to the information society. These general findings apply to both the technological and economic dimension of the information society.

The technological dimension

In general, Germany is undergoing a very dynamic information society development, but the pace varies considerably concerning different aspects. With regard to the spread and usage of ICT, Germany is still behind in international comparison, while in other aspects it ranks top internationally[12]. Lagging development can, in

[11] BBR (2005)
[12] BITKOM (2005)

particular, be observed regarding the use of modern ICT by private households. In contrast, German enterprises have recognized the advantages of ICT and use them intensively.

A rural-urban divide can be found in terms of ICT infrastructure supply. The availability and usage of broadband, for which there are rather high values in large parts of the country, also demonstrates that some peripheral and rural areas in central and north-eastern Germany have been left behind.

The economic dimension

The German ICT sector is constantly progressing, and the overall number of patents per year remains at a high level as well. While the regional distribution of high-tech patents again indicates an east-west division, data suggests a north-south-division of R&D expenditures and personnel, with several low R&D-oriented regions in the south, and very few regions with high R&D-orientation in northern Germany[13].

The social dimension

Regarding the usage of ICT within the education system, a much better infrastructure in schools is still needed in order to achieve an appropriate level of knowledge and technological competence. Regarding e-government, Germany is also lagging behind compared to much of Europe, with the related services so far being limited to providing access to information.

Policies supporting information society development

In Germany particular emphasis within the field of information society-related policies has been given – and is still given – to the provision of ICT infrastructures, although not to the same extent in all parts of Germany, as described above. Apart from pure infrastructure policies, special policy programmes concerning the development of the information society got underway at the federal level in the late 1990s[14]. These programmes aimed at combining actors from the political, economic, and social spheres in a joint partnership for innovation, and pursuing the overall objective of ensuring the participation of all social groups in the extensive use of ICT throughout Germany.

10.3 Information Society in a German peripheral Region – the Case of Tuttlingen

Within this project, a regional case study was conducted in the district of Tuttlingen. This district belongs to the federal state of Baden-Württemberg, a region that in terms of the economic strength of innovative functions ranks extremely highly in

[13] BITKOM (2005)

[14] EUROREG, IRS et al (2006)

national and even European comparison[15]. Thus, this region was chosen in order to gain some more detailed knowledge about successful development regarding the information society. At the same time, the district of Tuttlingen represents one of Germany's rural regions, which, generally speaking, have not been favoured by information society development so far. It was therefore expected to gain some insights and good practice examples useful for other rural or rather peripheral regions.

Although a rural region, the district of Tuttlingen is a very important location for the knowledge-intensive medical engineering industry, with employment figures as well as the gross value added in this field being above the national average in Germany[16]. The strong performance of Tuttlingen regarding innovative functions is indicated by the number of high-tech patents received in the region, which shows a strong regional cluster compared to other German areas.

This high performance within the knowledge-based economy is supported by a high level of education in the Tuttlingen district. Local and regional schools and professional training institutions offer appropriate educational careers within the local economy, and most of the medical engineering businesses interviewed recruit the main share of their employees from the region itself.

Regarding the availability of ICT, this case study region is very well equipped according to regional experts in the Tuttlingen area. With a few exceptions in some areas, broadband is available to businesses as well as private households. Not only is the availability quite high, but the usage as well. Especially businesses in the medical engineering sector need a high level of connectivity for their business relations and make regular use of the fast exchange of data.

Information society-related policy in Tuttlingen

The need to include areas that have not yet been connected to broadband infrastructures has recently been expressed, especially for new commercial areas, but also for remote villages. Local public authorities are currently lobbying for further development in this respect[17]. Their activities to improve the situation include informative events and discussions with relevant actors like providers and officials from all administrative levels in the region. At the same time, local policy makers, village mayors for instance, are given authority to find their own ways of supporting the development of an information society according to differentiated needs and local circumstances.

This differentiated inclusion of all levels in these processes is very important and has proved helpful for the success of information society-related development in this region. Specific local governance structures play quite an important role

[15] Statistik BW (2006)

[16] c.f. Meister-Scheuffelen (2006)

[17] c.f. doIT-regional (2006)

in this development. Close contacts between different actors in fields related to the information society help to mobilize support in overcoming obstacles along the way, especially when governance structures are characterized by strong local affinity.

10.4 Conclusions

An intensified use of ICT for communication results in a more effective workflow between different business locations, as well as between public administrations. These developments certainly have a strengthening effect on the cohesion level in these sectors. At the same time, fast access to information at all locations allows for a more sovereign, parallel working process within different units of administrations and businesses, thus fostering polycentric development as well. The possibilities of contacts between citizens and public administrations have changed as well, virtually diminishing distance.

The case study in the district of Tuttlingen illustrated a very successful example of the development of an information society in a rural region. As a result of this study, the following factors for enhancing an information society in this type of region were recognized:

(1) A stable economic situation: Because of the economic success of the region, public incomes are rather high. This indicates a necessity for the spread of modern ICT equipment in the public sector as well.

(2) The existence of a knowledge-based economy
- implies a higher need for ICT, thus fostering development in the business sector
- forms a stable basis for economic development and is bound to the region because of its specified knowledge (therefore fostering the regional educational level as well)

(3) Regional politicians as lobbyists, including their support through mediation processes with providers of ICT.

(4) Policies addressing the development of an information society.

These factors, which proved to have a strengthening effect on the development of the information society in the case study region, can be expected to have a positive influence on information society development in other rural regions as well.

All in all, some positive aspects of development were identified within this project, but also a few gaps and problems regarding the spatial development of the information society in Germany. Regarding the availability of ICT, the need was expressed to link the remaining areas to fast data connections. Furthermore, the constant pressure to keep up with the high speed of ICT development presents a challenge for all sectors of society. The growing overflow of information and lacking structures or linkages also pose additional difficulties in ICT use. These issues will need to be addressed in the future.

Furthermore, the spatial concentration of high-tech industries in southern German regions, as indicated by the aforementioned patent and R&D figures, implies that northern and eastern German regions have a need to further boost knowledge-intensive sectors in order to contribute to a balanced development of a knowledge-based economy, and thus to the development of an information society.

References

BITKOM (eds.) (2005) Daten zur Informationsgesellschaft. Status Quo und Perspektiven Deutschlands im internationalen Vergleich. Berlin

BBR (Federal Office for Building and Regional Planning) (ed.) (2005) Raumordnungsbericht 2005. Bonn

doIT-regional (Regional Policy Programme), Internet: www.doit-regional.de

EUROREG, IRS et al. (2006) Spatial aspects of the information society – Draft Final Report ESPON project 1.2.3., Internet: http://www.espon.eu/mmp/online/website/content/projects/259/654/file_2198/draft-fr_1.2.3_revised-full.pdf

EUROPEAN COMMISSION, eEUROPE (2005) An information society for all, Com(2002)/263 final

EUROPEAN COMMISSION (2006) ESDP – European Spatial Development Perspective. Potsdam

EUROPEAN UNION (2000) Lisbon Strategy, Internet: http://europa.eu/scadplus/glossary/lisbon_strategy_en.htm

MEISTER-SCHEUFELEN G (2006) Wirtschaftliche Perspektiven von Baden-Württemberg, Internet: www.statistik-bw.de

STATISTIK BW (Statistical Office of Baden-Wuerttemberg) (2006) *Statistisches Monatsheft Baden-Württemberg*, Issue 3/2006

WEFORUM (2004) Global Information Technology Report 2003-2004. Networked Readiness Index and Methodology. Internet: http://www.weforum.org/pdf/Gcr/GITR_2003_2004/Framework_Chapter.pdf

11 IRS Participates in EU-funded Research Project "Governance for Sustainability"

Eva Reisinger

Knowledge and learning have become key words in the discussion about the development of the knowledge-based society. Investing in knowledge and learning provides a fruitful basis for a society's future prospects through the increased innovation and creativity of its members. Knowledge and learning are therefore essential promoters of urban and regional development throughout Europe and thriving factors for future policies as well. Especially for the ambition of fostering sustainable development, knowledge and learning play a prominent role in better integrating environmental, social, and economic goals in policy-making. "Governance for Sustainability" (G-FORS), a Specific Targeted Research Project, funded through the Sixth EU Framework Programme, thus intends to draw attention to the role knowledge and learning are playing in policies of sustainability.

By conducting several case studies, G-FORS will explore the growing significance knowledge and learning play, especially concerning environmental politics. One focus of research is to analyze how newly adopted EU directives on issues of air quality management, environmental assessment, and emissions trading are implemented in different governance arrangements. In each of these policy fields the actual implementation follows to some extent a different governance mode, for example, in the case of emissions trading market mechanisms are especially crucial for its overall success. G-FORS is interested in studying how the concrete institutionalization and practical enactment of governance modes impact upon the effectiveness and legitimacy of these policies, and in enhancing our understanding of the role of knowledge and learning in these governance processes.

For that reason G-FORS will develop an innovative analytical model for the study of governance for sustainability, focusing on the synergy between new governance modes and different forms of knowledge, and taking into account rapid changes in the knowledge society. Researchers will identify a range of different forms of knowledge and analyse how these different forms (e.g., local, milieu, market, and expert knowledge) interact in the context of particular governance arrangements, while assuming that policies developed to promote sustainability will be biased and ineffective unless they effectively incorporate a full range of forms of knowledge. In particular the preliminary and ongoing work in Department 3 of the IRS regarding knowledge, knowledge development, and knowledge milieus linked to spatial development has given the G-FORS research team fundamental and indispensable clues and input.

G-FORS began in February 2006. For 36 months an organization of eleven partners from nine European countries, including two newer EU member states (Germany, Greece, Hungary, Italy, Netherlands, Norway, Poland, Sweden, and the United Kingdom), will work together to develop a conceptual framework for studying the above-mentioned topics. G-FORS will concentrate its case study work on urban areas. As more and more people in EU member states choose to live in cities, the development of sustainable policies is of vital importance within these urban areas and their surrounding regions. In cities, the heterogeneous range of actors (policy-makers, civil actors, officials and other representatives, experts, etc.), institutions, and governance arrangements should give rewarding insights which will help to gain a more legitimate understanding of sustainability. G-FORS hopes to assess how the engagement of a wider public can be activated to tackle any future threats of potential democratic deficits, and to further promote participation and sustainable development in the complex multi-level governance arrangements of the EU. The research team will develop a systematic strategy of disseminating research results to end-users at all levels of government and across Europe, whilst adding to our knowledge of the interaction between governance modes, knowledge, and broadened participation.[1]

[1] For more Information see: http://www.gfors.eu

Appendix

List of Authors

Prof. Dr. Joachim Burdack

was born in Berlin, Germany in 1951. He studied at the Free University of Berlin and at Indiana University (Bloomington, Indiana, USA). He received a Master of Arts (M.A.) at Indiana University, his Ph.D. (Dr. rer. nat.) at the Free University of Berlin in 1984, and a postdoctoral degree at Bamberg University in 1991. After working on projects in urban redevelopment in Berlin (West), he taught geography at Bamberg University from 1982 to 1993. Joachim Burdack joined the Leibniz Institute for Regional Geography in 1993 as a researcher. Since 2002 he has worked as the coordinator for urban and metropolitan studies. He became a professor of geography at Bamberg University in 1997 and at the University of Leipzig in 2000. His research interests include European metropolitan areas and the regional development of Western Europe.

Prof. Dr. Hans-Joachim Bürkner

was born in Hildesheim, Germany in 1954. After studying geography at the University of Göttingen he received his Ph.D in 1987. After employment as a scientific assistant he received his postdoctoral degree in 1994. After working as an assistant lecturer and extraordinary professor at the universities of Göttingen and Potsdam, as well as assistant and research coordinator at the Leibniz Institute for Regional Development and Structural Planning (IRS), he became a professor of economic and social geography at the University of Potsdam in 2002. Since then he has also been head of Research Department 4 (regeneration of shrinking towns and cities) at the IRS. His main research areas are urban development in Central and Southeast Europe, urban and regional development in border regions, and demographic change, migration and segregation in urban populations.

Mag. iur. Sylwia Czarnecka-Zawada

was born in Krakow, Poland in 1973. From 1992 to 1998 she studied law at the University of Krakow. From 2002 to 2003 she was the recipient of a scholarship from the Robert Bosch Foundation within a program for junior executives from Central and Eastern Europe. From 2003 to 2004 she was a postgraduate at the Technical University Dresden (LL.M.). Since 2004 she has been a scientific employee at the Leibniz Institute of Ecological and Regional Development.

Prof. Dr. Carl-Heinz David

was born in 1938. He passed the First Legal Exam in 1962 and the Bar Exam in 1967. He was the head of department and secretary general of the Zentralinstitut für Raumplanung at the University of Münster from 1968- 1975, has been the chair of Planning Law at the Faculty of Spatial Planning, University of Dortmund since 1975, and an Emeritus since September 2006. His research fields are spatial planning and environmental law, urban planning law, land law and policy, comparative studies in these fields (Great Britain, France, United States of America, Netherlands, Denmark, Sweden), and European law issues. He was the vice-president of the German Deans Conference for Architecture, Spatial Planning and Landscape Planning until 2006. In the 1990s he was a visiting professor at the Law Faculty at the University of Hawaii at Manoa, and in 2005 at the University of Basel.

Prof. Maroš Finka, Ph.D., M.Arch.

was born in Bratislava, Czechoslovakia in 1963. He has been a professor for Spatial and Landscape Planning at the Slovak University of Technolgy in Bratislava since 1996, and the director of the Central European Research and Training Centre in Spatial Planning, Centre of Excellence since 1999. His specialist areas are sustainable spatial planning, urban and cross-border physical planning, and strategic environmental assessment.

Dr. Hans-Peter Gatzweiler

was born in 1947. He studied geography, urban planning, and statistics and has been working in the field of spatial research and planning since 1972. His job focuses on analyses and forecasts of spatial and urban development in Germany and on related scientific and political advice. Since 1998 he has been the head of Department I (Spatial Planning and Urban Development) at the Federal Office for Building and Regional Planning (BBR), which is a higher government agency belonging to the Federal Ministry of Transport, Building and Urban Affairs (BMVBS). The department's task is to support both the BMVBS and other ministries in achieving their goals in the field of spatial planning and urban development.

PD Dr. Markus Hesse

was born in Arnsberg/Westfalen, Germany in 1960. He studied geography in Münster, got his Diploma in 1985, and his Ph.D. (dissertation in regional planning at the University of Dortmund, Faculty of Spatial Planning) in 1997. In 2004 he received a postdoctoral degree in human geography at the Free University of Berlin. From 1988-1998 he was a researcher at the Institute of Ecological Economic Research (IÖW) in Berlin, and from 1998 to 2004 a researcher at the Institute for Regional Development and Structural Planning (IRS) in Erkner (near Berlin). Since 2004 he has been a lecturer at the Free University of Berlin at the Institute of Geographical Sciences; Urban Studies Unit. His research areas include urban and regional development, suburbanization; transport, mobility and logistics in spatial contexts; and metropolitan studies.

Dr. Gerold Janssen

was born in 1962 and studied law in Göttingen. After completing the Legal Clerkship and the Bar Exam in 1994 he worked as a lawyer in Oldenburg. From 1997 to 1998 he worked as a research associate at the Baltic Institute for the Law of the Sea and Environmental Law (OSU) at the Law Faculty of Rostock University, where he received a Ph.D. in 2001 (his topic was the determination of protected marine areas). Since 1998 he has been a research associate at the Leibniz Institute of Ecological and Regional Development (IOER) in Dresden. From 2000 to 2001 he was head of the Strategies and Instruments for Resource-conscious Land Use research area, and from 2002 to 2005 the head of research in European Integration. He now also works as a lecturer at the law faculty of the Technical University Dresden. His activities are focused on international and European law, planning law, building law, and environmental law.

Dr. Rupert Kawka

was born 1969 in Freiburg im Breisgau, Germany. He received a Dipl.-Geograph, a Dipl.-Volkswirt and a Ph.D.. His thesis in geography was about African towns, and involved stays abroad in Nigeria and Burkina Faso. He has worked as a research fellow at the universities of Heidelberg and Frankfurt (1999-2000) and at the Halle Institute for Economic Research (IWH) (2001-2004). Since 2004 he has been a project coordinator at the Federal Office for Building and Regional Planning (BBR). He is a visiting lecturer for economic geography/regional analysis at the universities of Leipzig and Jena. His research projects and expert's reports, among other things, involve the regional costs of living, regional quality of life, rural regions and the new guiding principles for regional planning in Germany. He is a co-author of the German Spatial Development Report 2005.

Dipl.-Ing. Robert Knippschild

was born 1974. He studied spatial planning in Dortmund, Liverpool, Krakow, and Warsaw. Since 2001 he has been a research associate at the Leibniz Institute of Regional and Ecological Development (IOER) in Dresden. At the moment he is finalising his Ph.D. thesis on Cross-border Governance in Spatial Planning in the German-Polish-Czech Border Region. His fields of interest are communication and participation processes in spatial planning, cross-border co-operation in spatial planning, and cross-border strategy development. He has worked on various projects involving spatial research and policy advice, in particular along the new internal EU border of Eastern Germany.

Dipl.-Ing. Beate Konieczny

was born in Katowice, Poland in 1969. She studied spatial planning at the University of Dortmund in Germany and at the Gdansk University of Technology in Poland. In her diploma thesis she dealt with cross-border co-operation as a chance for urban development in consideration of the regional housing market situation in the German-Polish twin-city Guben-Gubin. She is now a scientific assistant and Ph.D. student at the Leibniz-Institute of Ecological and Regional Development (IOER) in Dresden. Her current project is aimed at creating a strategic concept for coordinated spatial planning in the German-Polish Odra Estuary Region and developing recommendations for the organization of cross-border Integrated Coastal Zone Management (ICZM). The topic of her Ph.D. dissertation is institutional capacity building in German-Polish cross-border regions, with particular attention paid to the role of spatial and environmental planning.

Dr. Markus Leibenath

was born in Detmold, Germany in 1966. He studied landscape ecology, spatial planning and resource economics at the Technical University München and Harvard University. He received his Ph.D. at the Technical University Berlin with a thesis on nature conservation and regional development. Since 2000 he has worked as a senior researcher at the Leibniz Institute of Ecological and Regional Development (IOER) in Dresden. His work centres on transnational spatial planning and on cross-border aspects of biodiversity and landscape policies.

Prof. Dr. Ulf Matthiesen

studied sociology, philosophy and law in Freiburg, Frankfurt/Main and Paris. His postdoctoral degree was in sociology at the J. W. Goethe University in Frankfurt/Main. He has taught and researched at various universities and within different project-networks (Dortmund-VW-Project, Frankfurt/Main, Erlangen). He is presently the head of research at Department 3 (Knowledge Milieus and Spatial Structures) at the Leibniz Institute for Regional Development and Structural Planning (IRS) in Erkner and a professor at the Institute for European Ethnology (Humboldt University) in Berlin. His main areas of research at present are co-evolution of knowledge and space; knowledge-based city-regional development, especially knowledge-milieus and *KnowledgeScapes*; suburbia development and suburbanization processes; cultural landscapes and *Space Pioneers*; governance modes and knowledge forms; regional culinaric systems in Europe and their developmental potential (food sector, tourism, identity etc.).

Dipl.-Geogr. Kristine Müller

was born in Schwedt/Oder, Germany in 1975. After highschool and a year abroad in the USA, she studied geography, sociology and political science at the Humboldt-University Berlin, and at the U.D.C. in Washington (USA) and the UvA in Amsterdam, NL. Her main areas of interest include research on migration and inequality, transformation processes, and border research. After completing her studies she worked as a research assistant at the Institute for Research on Social Chances (ISO-Institute) in Cologne. Since October 2005 she has been working at the Institute for Regional Development and Structural Planning (IRS) in Erkner. There she began her research activities within projects of the European Spatial Planning and Observation Network (ESPON), as well as a study on transnational co-operation in the German-Polish border regions. Research on borders within the enlarged European Union will be in the main focus of work from 2007 onwards.

Mag. phil. Eva Reisinger

was born in Vienna, Austria in 1972. She studied geography, with a main focus on economic geography. Her activities have also included gender research in regional development at the University of Vienna. From 2003 to 2005 she was involved in a research and teaching project concerning gender studies in geosciences. Since 2006 she has been a scientific employee. Her research interests include urban and regional develop-ment, sustainability research, qualitative methods of empirical social research, fem-inist geography, and geography sciences and technologies.

Prof. Dr.-Ing. Dietmar Scholich

was born in 1947. He studied civil engineering (he became a graduate engineer in 1973) and spatial and environmental planning (he became a graduate engineer in 1976). In 1982 he finished his Ph.D. After working as a public employee in Kaiserslautern (Garten- und Grünflächenamt) and Oldenburg (Amt für Straßen und Brücken, Planungsamt), he became the Secretary General at the Academy for Spatial Research and Planning (ARL). Since 2000 he has been an honorary professor in land and regional planning at the Faculty of Architecture, Slovak Technical University Bratislava.

Dipl.-Ing. Lars Stratmann

was born in Bremerhaven, Germany in 1975 and studied landscape architecture, (his main emphasis was on landscape planning) at the Technical University Dresden. He then worked for Plan T – Planning Group Landscape and Environment in Dresden and Radebeul, where his main focus was on landscape planning and flood protection concepts/measures. He is now a scientific employee at the Leibniz Institute of Ecological and Regional Development (IOER) in Dresden, where his main activities involve the instruments of environmental planning and the management of compensation measures on a regional level. In his contribution to a book about European Groupings of Territorial Co-operation he reported about his experience with the INTERREG III A-project, which concerned strategic environmental assessment for regional planning in borderland regions.

Prof. Dr. Wendelin Strubelt

was born in 1943. He studied protestant theology, political science, history and German literature. He graduated in 1969 with an M.A. (political science and German literature). In 1970 he became a research fellow in the Department of Political Science at the University of Konstanz. In 1973 he became a lecturer in social science at the University of Bremen and received his Ph.D. in political science and sociology in 1976. In 1977 he became an assistant professor in the Department of Social Science at the University of Bremen. In 1979 he became an associate professor at the University of Bremen. Since 1981 he has been the director and a professor at the Federal Research Institute for Regional Geography and Regional Planning. Since 1998 he has been the vice-president and professor at the Federal Office for Building and Regional Planning (BBR).

Dr. Sabine Zillmer

was born in 1968 and has an academic background in economics, acquired at Göttingen University and the Flinders University of South Australia. For her Ph.D. within an international research network at the Humboldt University of Berlin, she analysed the labour supply behaviour of Polish agricultural households. In 2002 she joined the Leibniz Institute for Regional Development and Structural Planning (IRS) as a research fellow and has since worked on several projects. Among them are evaluations of PHARE CBC and INTERREG programmes at the German-Polish border. She is also involved in the co-ordination and methodological and analytical work of different ESPON projects. In addition, she has gathered knowledge-economy related research experience in the institute's research projects, for example, "The Impact of the Service Economy on Metropolitan Regions: Organisation, Mobility, and Communication", "Urban Regions near Metropolis - Their Role in the Network of Metropolitan Information Services", and "Implications of the Knowledge Economy for the Development of the European City System".

Appendix

Portrait of the Editors Instituitions

Academy for Spatial Research and Planning (ARL)

The ARL is concerned with research about spatial structures and developments that are important for Germany, and of international relevance. The ARL is a service institution for basic and application-oriented research. At the same time, it is a neutral forum which promotes discourse about spatial science issues. The ARL network currently consists of approximately 1,000 experts. Their tasks are, among other things, to initiate research and to perform joint research within this network, as well as to encourage a transfer of knowledge.

Academy for Spatial Research and Planning (ARL)
Hohenzollernstr. 11
30161 Hanover

E-Mail: ARL@ARL-net.de
Internet: www.ARL-net.de
Secretary General: Prof. Dr-Ing. Dietmar Scholich

Leibniz Institute for Regional Geography (IfL)

The Leibniz Institute for Regional Geography (IfL) conducts basic research into the regional geography of Germany and Europe and provides – as one part of its educational task – geographical information about spatial structures and their development. Additional research topics are the analysis of spatial structures and development on different scales, as well as the presentation of the geography of Germany abroad.

Leibniz Institute for Regional Geography (IfL)
Schongauerstr. 9
04329 Leipzig

E-Mail: info@ifl-leipzig.de
Internet: www.ifl-leipzig.de
Director: Prof. Dr. Sebastian Lentz

Leibniz Institute of Ecological and Regional Development (IOER)

The Leibniz Institute of Ecological and Regional Development (IOER) carries out interdisciplinary research into the complex interaction of the natural, built and, social environments. Key research areas are the environmental aspects of regional development on an European level, strategies for the ecological restructuring of towns and regions, the environmental impact of land use changes, as well as the material flows of building and housing.

Leibniz Institute of Ecological and Regional Development (IOER)
Weberplatz 1
01217 Dresden

E-Mail: info@ioer.de
Internet: www.ioer.de
Director: Prof. Dr. Bernhard Müller

Leibniz Institute for Regional Development and Structural Planning (IRS)

The Leibniz Institute for Regional Development and Structural Planning (IRS) in Erkner (near Berlin) examines the spatial, social, and economic basics and planning perspectives for regions and towns. Research conducted by the IRS deals with the following topics: strategies for regionalisation, institutional change, regional structures and the restructuring of towns, as well as knowledge milieus in the context of transformation processes.

Leibniz Institute for Regional Development and Structural Planning (IRS)
Flakenstr. 28-31
15537 Erkner

E-Mail: regional@irs-net.de
Internet: www.irs-net.de
Director: Prof. Dr. Heiderose Kilper

Federal Office for Building and Regional Planning (BBR)

The Federal Office for Building and Regional Planning is a superior federal authority in the operational division of the Federal Ministry of Transport, Building and Urban Affairs (BMVBS). On the one hand, the BBR supports the federal government, with sectoral scientific consultation in the policy areas of spatial planning, urban development, housing and building. On the other hand, it is responsible for the supervision of Germany's most important federal buildings at home and abroad.

Federal Office for Building and Regional Planning (BBR)
Deichmanns Aue 31-37
53179 Bonn

Fasanenstr. 87
10623 Berlin

E-Mail: zentrale@bbr.bund.de
Internet: www.bbr.bund.de
President: Dipl.-Ing. Florian Mausbach
Vice President: Prof. Dr. Wendelin Strubelt

Printing: Krips bv, Meppel, The Netherlands
Binding: Stürtz, Würzburg, Germany